Vol. 29. **The Analytical Chemistry of Sulfur and Its Compounds** (*in three parts*). By J. H. Karchmer

Vol. 30. **Ultramicro Elemental Analysis.** By Günther Tölg

Vol. 31. **Photometric Organic Analysis** (*in two parts*). By Eugene Sawicki

Vol. 32. **Determination of Organic Compounds: Methods and Procedures.** By Frederick T. Weiss

Vol. 33. **Masking and Demasking of Chemical Reactions.** By D. D. Perrin

Vol. 34. **Neutron Activation Analysis.** By D. De Soete, R. Gijbels, and J. Hoste

Vol. 35. **Laser Raman Spectroscopy.** By Marvin C. Tobin

Vol. 36. **Emission Spectrochemical Analysis.** By Morris Slavin

Vol. 37. **Analytical Chemistry of Phosphorus Compounds.** Edited by M. Halmann

Vol. 38. **Luminescence Spectrometry in Analytical Chemistry.** By J. D. Winefordner, S. G. Schulman and T. C. O'Haver

Vol. 39. **Activation Analysis with Neutron Generators.** By Sam S. Nargolwalla and Edwin P. Przybylowicz

Vol. 40. **Determination of Gaseous Elements in Metals.** Edited by Lynn L. Lewis, Laben M. Melnick, and Ben D. Holt

Vol. 41. **Analysis of Silicones.** Edited by A. Lee Smith

Vol. 42. **Foundations of Ultracentrifugal Analysis.** By H. Fujita

Vol. 43. **Chemical Infrared Fourier Transform Spectroscopy.** By Peter R. Griffiths

Vol. 44. **Microscale Manipulations in Chemistry.** By T. S. Ma and V. Horak

Vol. 45. **Thermometric Titrations.** By J. Barthel

Vol. 46. **Trace Analysis: Spectroscopic Methods for Elements.** Edited by J. D. Winefordner

Vol. 47. **Contamination Control in Trace Element Analysis.** By Morris Zief and James W. Mitchell

Vol. 48. **Analytical Applications of NMR.** By D. E. Leyden and R. H. Cox

Vol. 49. **Measurement of Dissolved Oxygen.** By Michael L. Hitchman

Vol. 50. **Analytical Laser Spectroscopy.** Edited by Nicolo Omenetto

Vol. 51. **Trace Element Analysis of Geological Materials.** By Roger D. Reeves and Robert R. Brooks

Vol. 52. **Chemical Analysis by Microwave Rotational Spectroscopy.** By Ravi Varma and Lawrence W. Hrubesh

Vol. 53. **Information Theory As Applied to Chemical Analysis.** By Karel Eckschlager and Vladimir Štěpánek

Vol. 54. **Applied Infrared Spectroscopy: Fundamentals, Techniques, and Analytical Problem-solving.** By A. Lee Smith

Vol. 55. **Archaeological Chemistry.** By Zvi Goffer

Vol. 56. **Immobilized Enzymes in Analytical and Clinical Chemistry.** By P. W. Carr and L. D. Bowers

Vol. 57. **Photoacoustics and Photoacoustic Spectroscopy.** By Allan Rosencwaig

Vol. 58. **Analysis of Pesticide Residues.** Edited by H. Anson Moye

Vol. 59. **Affinity Chromatography.** By William H. Scouten

Vol. 60. **Quality Control in Analytical Chemistry.** By G. Kateman and F. W. Pijpers

Vol. 61. **Direct Characterization of Fineparticles.** By Brian H. Kaye

Vol. 62. **Flow Injection Analysis.** By J. Ruzicka and E. H. Hansen

Planar Chromatography in the Life Sciences

CHEMICAL ANALYSIS

A SERIES OF MONOGRAPHS ON
ANALYTICAL CHEMISTRY AND ITS APPLICATIONS

VOLUME 108

A WILEY-INTERSCIENCE PUBLICATION

JOHN WILEY & SONS

New York / Chichester / Brisbane / Toronto / Singapore

Planar Chromatography in the Life Sciences

Edited by

JOSEPH C. TOUCHSTONE

University of Pennsylvania
School of Medicine
Philadelphia, Pennsylvania

A WILEY-INTERSCIENCE PUBLICATION

JOHN WILEY & SONS

New York / Chichester / Brisbane / Toronto / Singapore

Library of Congress Cataloging in Publication Data:
Planar chromatography in the life sciences/[edited by] Joseph C.
Touchstone
p. cm. — (Chemical analysis; v. 108)
"A Wiley-Interscience publication."
Bibliography: p.
ISBN 0-471-50109-3
1. Thin layer chromatography. 2. Biochemistry—Technique.
I. Touchstone, Joseph C. II. Series.
QP519.9.T55P53 1989
574.19′285—dc20 89-5748 CIP

Printed in the United States of America
10 9 8 7 6 5 4 3 2 1

CONTRIBUTORS

Juan G. Alvarez, School of Medicine, University of Pennsylvania, Philadelphia, PA 19104

Daniel W. Armstrong, Department of Chemistry, University of Missouri, Rolla, MO 65401

John A. Baino, Merck Sharp and Dohme, Rahway, NJ 07065

Gary Bicker, Merck Sharp and Dohme, Rahway, NJ 07065

Edward T. Butts, Whatman, Inc., Clifton, NJ 07014

Fiona M. Clark, Merck Sharp and Dohme, Rahway, NJ 07065

David Y. Cooper, School of Medicine, University of Pennsylvania, Philadelphia, PA 19104

Dean Ellison, Merck Sharp and Dohme, Rahway, NJ 07065

Thomas Enzweiler, EM Science, Cherry Hill, NJ 08034

Heinz Filthuth, Berthold, Wildbad, Germany

Stanley Fowler, School of Medicine, University of South Carolina, Columbia, SC 29208

Martin Gould, EM Science, Cherry Hill, NJ 08034

Nelu Grinberg, Research Laboratories, Merck Sharp and Dohme, Rahway, NJ 07065

Soon M. Han, Department of Chemistry, University of Missouri, Rolla, MO 65401

Elaine Heilweil, Whatman Inc., Clifton, NJ 07014

James A. Herman, Analect Instruments, Irvine, CA 92714

Eric J. Levin, School of Medicine, University of Pennsylvania, Philadelphia, PA 19104

Sidney S. Levin, School of Medicine, University of Pennsylvania, Philadelphia, PA 19104

Edward Rapkin, IN/US Service Corporation, Fairfield, NJ 07006

Warren E. Schwartz, Whatman, Inc., Clifton, NJ 07014

Megumi Saito, Medical College of Virginia, Virginia Commonwealth University, Richmond, VA 23298

Dean E. Sequera, Shimadzui, Inc., Columbia, MD 21046

Kenneth H. Shafer, Analect Instruments, Irvine, CA 92714

Bayard T. Storey, School of Medicine, University of Pennsylvania, Philadelphia, PA 19104

Joseph C. Touchstone, School of Medicine, University of Pennsylvania, Philadelphia, PA 19104

Pat Tway, Merck Sharp and Dohme, Rahway, NJ 07065

Tom R. Watkins, Department of Pediatrics, School of Medicine, New York University, New York, NY 10016

Robert K. Yu, Medical College of Virginia, Virginia Commonwealth University, Richmond, VA 23298

PREFACE

Planar chromatography (PC) or thin-layer chromatography (TLC) has progressed considerably in recent years. There have been a number of symposia and meetings devoted to this subject, particularly in Europe. Because of the heightened interest it was felt that a need existed for presentation of up-to-date information concerning recent developments in the field, particularly as they related to biochemical parameters. Concurrently with developments in the chromatographic aspects there has been an interest on the part of the manufacturers to provide scanners for evaluation of the separations. This includes scanners for spectrocharacterization as well as quantitation of isotope levels. The symposium was organized with an aim to provide some presentations in both areas. The symposium appeared very successful: Thus it seemed appropriate to collect the presentations in a volume to enable others to obtain some update to what has been happening in PC (TLC). This meeting would not have been possible without the support of the Federation of the Societies for Experimental Biology and Medicine. As a satellite symposium, exposure to a greater audience than otherwise would have been possible resulted. The support of the suppliers or manufacturers who participated is gratefully acknowledged.

In addition, we are grateful to the speakers for forwarding their manuscripts which form the basis of this volume.

JOSEPH C. TOUCHSTONE

Philadelphia, Pennsylvania
November 1989

CONTENTS

CHAPTER 1	**CURRENTS IN PLANAR CHROMATOGRAPHY** *Joseph C. Touchstone*	1
CHAPTER 2	**RAPID DETECTION AND QUANTITATION OF LIPIDS ON THIN-LAYER CHROMATOGRAPHY BY NILE RED FLUORESCENCE** *Stanley D. Fowler*	7
CHAPTER 3	**USE OF BONDED PHASES IN PLANAR CHROMATOGRAPHY** *Martin Gould and Thomas Enzweiler*	15
CHAPTER 4	**ANALYSIS OF BUTYRIC ACID IN A MODEL SYSTEM** *Elaine Heilweil, Edward T. Butts, Fiona M. Clark, and Warren E. Schwartz*	49
CHAPTER 5	**TLC-IMMUNOSTAINING OF GLYCOLIPIDS** *Megumi Saito and Robert K. Yu*	59
CHAPTER 6	**TLC IN PHARMACEUTICAL RESEARCH** *Nelu Grinberg, John A. Baino, Gary Bicker, Patricia Tway, and Dean Ellison*	69
CHAPTER 7	**ENANTIOMERIC SEPARATION BY THIN-LAYER CHROMATOGRAPHY** *Soon M. Han and Daniel W. Armstrong*	81
CHAPTER 8	**ASSAY OF BIPHENYL METABOLITES BY HPTLC-SPECTRODENSITOMETRY** *Sidney S. Levin, Joseph C. Touchstone, and David Y. Cooper*	101

CHAPTER 9 IN SITU DETERMINATION OF MALONDIALDEHYDE ON THIN-LAYER PLATES BY FLUORESCENCE SPECTRODENSITOMETRY 111
Juan G. Alvarez, Bayard T. Storey, and Joseph C. Touchstone

CHAPTER 10 ANALYSIS OF ASCORBIC ACID BY THIN-LAYER CHROMATOGRAPHY 119
Joseph C. Touchstone, Tom R. Watkins, and Eric J. Levin

CHAPTER 11 ONE- AND TWO-DIMENSIONAL SCANNING FOR ^{32}P AND OTHER UNCOMMON TAGS 127
Edward Rapkin

CHAPTER 12 BIOANALYTICAL APPLICATION OF THIN-LAYER CHROMATOGRAPHY/FOURIER TRANSFORM INFRARED SPECTROMETRY 157
James A. Herman and Kenneth H. Shafer

CHAPTER 13 DETECTION OF RADIOACTIVITY DISTRIBUTION WITH POSITION-SENSITIVE DETECTORS, LINEAR ANALYZER, AND DIGITAL AUTORADIOGRAPH 167
Heinz Filthuth

CHAPTER 14 NEW TECHNIQUES IN TWO-DIMENSIONAL DATA PROCESSING 185
Dean E. Sequera

INDEX 197

Planar Chromatography in the Life Sciences

CHAPTER

1

CURRENTS IN PLANAR CHROMATOGRAPHY

JOSEPH C. TOUCHSTONE

Planar chromatography (PC), a term only recently coined as perhaps a better designation for thin-layer chromatography (TLC), remains a viable technique. It is subject to continual improvement not only in the sorbents available but in the instrumentation available to facilitate its use. Advances in production of stationary phases have contributed greatly to advances in PC. This contribution, spurred by the continual development of sorbents for use in high-performance liquid chromatography, may be the major factor in growth of PC. At present there are a large number of stationary phases available for use in thin layers. These include normal as well as reversed phases, ion exchange, and the gel-type layers. Largely due to the efforts of the manufacturers of the layers, it is now possible to purchase reproducible layers for continual use. Few practitioners now prepare their own layers. More recently the advent of 5-μm sorbents of reproducible quality have become available. The so-called high-performance TLC (HPTLC) has seen increasing growth as chromatographers begin to realize that the use of HPTLC provides a sensitive and reproducible analytical method. These layers are smaller and faster. Smaller samples and better detection limits result. Considering that multiple samples can be separated simultaneously in a PC system whether conventional or high performance, this methodology is much more economical than other chromatographic methods. There have been some attempts at automating the PC. At this writing, there is no fully automated planar chromatographic system. Considering that it appears to be possible to assay 3×10^6 samples using one instrument (1), it seems that automated PC is not far off. Certain modules of a possible automatic system are presently available.

Currently, increasing numbers of investigators are using preparative PC. Perhaps the centrifugally accelerated preparative PC methodology may provide viable alternatives to other separation methods (2). This instrumentation is available and the layers for use in the process can be had in a variety of modes.

Developments in planar chromatography are for the most part limited to what has become available in the sorbents ready for use. Some of the more recent modes of thin layers include cellulosic sorbents that have been coated as layers. These resemble paper chromatography in many properties. The plates are more stable, give decreased development times, and in many cases show better resolution and sensitivity. These layers give good results with polar compounds such as carboxylic acids and carbohydrates. Celluloses have been modified with ion exchangers, polyethyleneimine, and silica gel for special purposes. Bonded phases may at this point have replaced unmodified silica gel for special purposes and as the most often used sorbent. Many of these phases have resulted from developments in HPLC. These phases can be either polar or non-polar depending on the substituent used for modifying the sorbent. Many of the newly developed sorbents and the applications for which they were developed are described in Chapter 3. There is such a wide variety of modifications now available that it is often possible to fit a layer to the separation problem at hand.

Before chromatography can be carried out some attention must be focused on sample preparation. There are a number of companies that provide solid-phase extraction (SPE) devices for sample preparation. Sample preparation can be the most important step in all forms of chromatography. For this reason, there has been a rapid gain in acceptance of the numerous SPE devices. They provide an efficient, reproducible, and cost-effective sample preparation method. With proper usage of SPE, the methodology can be tailored to selectively separate the analyte or to remove the analyte of choice if proper manipulation of the eluents and their sequence of use is determined.

SPE technology can be utilized with sorbents that fall into the classes below:

1. *Normal Phase.* Usually silica and separation depends on polar interactions,
2. *Reversed Phase.* Usually modified silica and relies on hydrophobic interactions,
3. *Ion Exchange.* Generally of resin type but includes some modified silica; basic separation mechanism is ionic interaction,
4. *Size exclusion.* Based on Sephadex or gel-type sorbents; separation on base of molecular size or steric interactions.

The SPE can be used in three types of modes:

1. *Analyte Retention.* The mobile phases or eluents are tailored to allow the impurities to pass through. This will vary with the sorbent type.
2. *Sample Concentration.* Large volumes of samples are passed through the columns. This must be set up to retain the analyte, which is then

eluted with a small volume of solvent. Typically, a litre of wastewater can be extracted to yield the analyte in a 0.2 to 1 mL volume of eluent.

3. The method can be tailored to have the impurities retained and the analyte passed through for collection.

These methods generally follow four steps in the process:

1. The column is "conditioned" by passing through it a series of solvents. These usually are based on the eluents that may be used during the extraction,
2. The sample is added, usually in solution, sometimes the original sample which has been adjusted for pH or other requirement,
3. The column is washed to remove the interfering substances, the "junk" that the analyst wishes to remove,
4. The compound of interest is selectively eluted in as small a volume as possible to permit complete recovery.

This equipment, perhaps with some modification, can also be extended to apply the sample or an aliquot of it to a TLC layer. No claims are made as to the completeness of this information. The advent of SPE may be one of the most important events in the progress of the use of chromatography in the analytical process.

Also available are several specialized instruments which may, when perfected, aid in further development in growth of planar chromatography.

1. Overpressured thin-layer chromatography embodied in the "Chrompres" produced by Hungarian Manufacturers has not seen wide use in this country. With some refinements and more ease of use, perhaps it may find a wider market. At present, it has greater popularity in European laboratories. Numerous papers with applications have appeared mostly in European journals. The process involves applying pressure to the surface of the layer using an inflatible membrane. The pressure can be provided by compressed air or water. The mobile phase is then pumped through the layer. Shorter development time and less solvent use follow.
2. In a combination of the major features of the overpressured concept, HPTLC and GPLC, the Hyltex instrument for high-pressure planar chromatography (Heyltex Corp., Houston, TX) has several advantages. It may be considered a modification of the Camag U-Chamber. Separations can be accomplished in only 1–2 min. As many as 1–50

samples can be processed using only 1 mL of mobile phase which is fed under pressure. The pressure upon the layer is provided by a hydraulic press.

3. Centrifugal chromatography, as embodied in the Chromatotron (Harrison Research Palo Alto, CA), was developed to accelerate separation on TLC for preparative purposes. Circular layers for use with this instrument are presently being produced by Analtech Inc., Newark, DE (see listing in sorbent suppliers). Rapid separations (20 min) can be accomplished and the components collected without scraping of the layer. Sorbent layers can be regenerated and the separation repeated. As much as 1 g of sample can be separated.

These instruments, for the most part, have not been generally adopted. When they are perfected, they should see much wider usage. Considering their speed and low solvent consumption, these devices should attract more attention. The greatest advances in instrumentation for planar chromatography are the more recent improvements in scanners for both absorbance and radiation assessment. Thin-layer chromatography and its companion high-performance thin-layer chromatography are recognized as viable techniques for many analytical problems. The advantages lie in high sample throughput and cost effectiveness. The present availability of scanning densitometers (as well as imaging scanners) and isotope scanners has made the technique a highly efficient quantitative tool. Scanners typically can detect UV or visible absorbance and fluorescence. The use of fluorescene measurement has enabled densitometry of TLC or HPTLC to become one of the most sensitive quantitative methods. Some instruments primarily designed for gel scanning are also being used for TLC. It is now possible to scan the absorbance spectrum of a substance separated on a thin layer without changing or removing it from the layer, as described in discussions later in this volume. There are more than fifteen companies now providing imaging scanners. These include optical as well as video scanning devices. With rapid advances being made in detection elements it seems possible that more use will be made of electronic scanning.

Isotopic labeling of compounds and their metabolites in biological systems as well as a number of other applications have become an increasingly popular subject. Scanning paper chromatograms and TLC were used often during the decade of 1960–1970. Interest waned during 1970–1980 due to the lack of suitable instrumentation. New intrument concepts, particularly in the detection devices, during the 1980s provided greater accuracy along with 2D imaging. It became possible to scan ^{32}P. At present, the instrumentation for scanning TLC directly may compete with autoradiography. Certainly, it may forewarn the demise of scintillation counting with its time-consuming scraping of the zones from a TLC plate. A major problem in the research laboratory involved with TLC of radiolabeled compounds is in detection and quantitation. Autoradiography and liquid scintillation counting are still widely used but both have limitations. Autoradiography can be time-consuming and usually does not provide quantitative data. Liquid scintillation methods are also time-consuming, require numerous vials, and, because of the radioactivity, can be a safety hazard. To enhance the TLC analysis of radiolabeled metabolites usually encountered in biomedical research, scanners are now available that provide the sensitivity and spatial analysis that is required. Imaging scanners provide a 100-fold improvement in sensitivity over mechanical scanners. They are much faster (minutes not days) because no scraping or vials are required. HPTLC layers having layers thinner than the conventional 250-μm layer will count at a higher efficiency. Counting efficiencies are uniform over the entire layer. The graphic displays provided by the scanners are dramatic. Prompt visualization of separated radioactivity and ready localization of the zones results.

With the advent of more sophisticated instrumentation and capabilities to assay for ^{35}S, ^{32}P, and ^{125}I, those using TLC can now scan with much more reproducible and sensitive results. These developments will do much to further the advance of tracer studies in the biomedical and other areas.

Planar chromatography will continue to advance if indications of this review are projections of the future. At present, although some steps are automated, no apparatus fully automating the entire process is yet available in a single unit. When this is ready the potential to assay and process 10×10^6 samples per year appears to be in reach. It should provide a highly favorable, cost–effective assay method (1). With highly specialized programs for operation on the instruments by computer, automation should not be far off. The following papers, which resulted from the symposium, have been organized into two sections: (*a*) reports concerning methods for separation and final quantitation with emphasis on densitometry, and (*b*) descriptions of some of the more recent developments in both densitometry and isotope scanning.

REFERENCES

1. D. Rogers, *Am. Lab.* 16: 65 (1984).
2. R. J. Laub and D. L. Zink, *Am. Lab.* 13: 55 (1981).

CHAPTER

2

RAPID DETECTION AND QUANTITATION OF LIPIDS ON THIN-LAYER CHROMATOGRAPHY BY NILE RED FLUORESCENCE

STANLEY D. FOWLER

The remarkable pace of research in biochemistry of the past two decades has been aided by the wide application of simple techniques to survey changing patterns of major biological constituents in tissues. Two important examples are the electrophoretic separation of proteins and nucleic acids on matrix gels and their detection, respectively, with commassi blue and ethidium bromide. For analysis of lipids, thin-layer chromatography (TLC) ranks as the corresponding technique of choice. The near infinite mixture of solvents, the wide variety of separation media available, and the choice of one- or two-dimensional protocols permit rapid separation of virtually any class of lipids and many subclass mixtures of lipids with differing fatty acid moieties.

Limiting in TLC, however, is the availability of a good universal reagent to detect and quantitate the lipids once they are separated. This reagent should be easily applied, very sensitive, and non destructive, and should allow additional experimental manipulations such as radioisotope counting or solvent extraction of selected bands for purification. Fluorescent dyes best fit this description, but most reported so far for use with TLC are significantly limited by rapid fading (1). Recently, however, we found an intensely fluorescent, photostable dye, *nile red*, which can be used as a universal stain for the detection and in situ quantitation of lipids on TLC plates (2). The dye selectively stains only lipids or other hydrophobic compounds such as drugs. The versatility of TLC combined with nile red staining now give the lipid chemist the same advantages of flexibility and rapid analysis enjoyed by protein and nucleic acid researchers who use electrophoretic methods.

PROPERTIES OF NILE RED

Nile red is an uncharged benzophenoxazone dye (Figure 2.1) derived from the oxidation of nile blue, a dye commonly employed as a histochemical stain

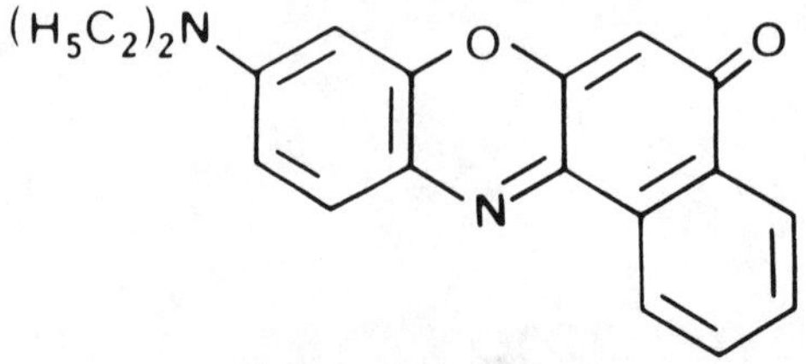

NILE RED

Figure 2.1. Nile red (9-diethylamino-5*H*-benzo[α]phenoxazine-5-one) is a fluorescent hydrophobic probe.

for fat. Nile red possesses four properties that make it a useful and sensitive lipid stain (3, 4). First, the fluorescene of nile red is very strong, approaching the intense fluorescence of rhodamine B. Second, the dye fluoresces only in an hydrophobic environment. Its excitation and emission maximum shift, however, depending on the dielectric constant of the solvent in which it is dissolved. Third, nile red is photochemically stable fluorochrome; it has been employed for that purpose in dye lasers. Fourth, nile red is quite soluble in organic solvents, but only sparingly soluble in water; the partition coefficient of nile red in organic solvents relative to water is approximately 200. As a result of these properties, nile red should both preferentially dissolve and fluoresce in lipid, providing selectivity as a lipid stain. Further, even small amounts of lipid should be detectable from the strong fluorescene of dissolved dye. And, the staining produced on TLC plates should be sufficiently stable to allow quantitation of the fluorescent lipid bands present before fading occurs.

MICROASSAY FOR DETECTION AND QUANTITATION OF LIPIDS ON TLC PLATES USING NILE RED

A simple procedure is outlined in Figure 2.2 for staining lipids and other hydrophobic compounds with nile red.* We found that brief dipping of the TLC plate in a tank of dye provided more uniform staining than did spraying dye on the plate with an atomizer. Immediately after staining, all lipids

* Nile red can be obtained commercially from Kodak Laboratory & Specialty Chemicals (Rochester, NY), Molecular Probes, Inc. (Junction City, OR), Polysciences, Inc. (Warrington, PA), and Lambda Physik GmbH (Gottingen, Federal Republic of Germany).

Following separation of lipids on silica gel TLC plates (Whatman type 60A, K6):

1. Stain plate by dipping it 5 s in a tank of nile red solution (8 μg/mL methanol–water 80:20, v/v). Dry plate 5 min at 100 °C
2. Decolorize plate by dipping it 5 s in a tank of dilute bleach (Clorox 1/2500). Dry plate 30 min in vacuum oven at 100 °C.
3. Visualize fluorescent bands with short-wavelength ultra-violet lamp. Quantitate fluorescence by in situ fluorometry (Ex/Em 580/640 nm).

Figure 2.2. Detection and quantitation of lipids on thin-layer chromatograms by nile red staining.

fluoresce brilliantly. But as the plate dries, a strong background fluorescence appears over the entire plate. This background fluorescence can be removed by misting the plate with water, and it will also gradually disappear if the stained plate is exposed to sun light. However, a simple and permanent means of removing the background fluorescence is to treat the stained plate with very dilute bleach (Figure 2.3). With this treatment, dye adsorbed onto silica gel is oxidized to a nonflorescent product while the dye dissolved in lipids is spared and continues to fluoresce.

As shown in Figure 2.3, both neutral and phospholipids are detected with the procedure. Since nile red is an hydrophobic probe (3, 4), we anticipated that the color of the dye would vary between yellow and red, depending on the lipid class stained. Surprisingly, however, a pink fluorescence is seen regardless of lipid type stained. The visual impression is confirmed by studies of fluorescene spectra of individual stained bands on the TLC plate, which show a single common emission peaking at 640 nm. Emission of a single fluorescent color is fortunate, for it permits quantitation of fluorescence intensity of stained bands by a single measurement at one wavelength pair (excitation, 580 nm; emission 640 nm). Use of a fluorometer with TLC scanning attachment to measure the nile red florescence allows rapid in situ quantitation of lipids on thin-layer chromatograms.

The conditions given in Figure 2.2 have been optimized to give maximal fluorescence staining and reproducibility. The silica gel TLC plates are prewashed in chloform–methanol 2:1 (v/v) prior to use to reduce background, and narrow sorbent lanes (5 mm wide) are etched on the plate to limit expansion of sample spot diameter during chromatography (5). Typical mass calibration plots generated from scanning stained TLC plates with a fluorometer are shown in Figure 2.4 for both neutral lipids and phospholi-

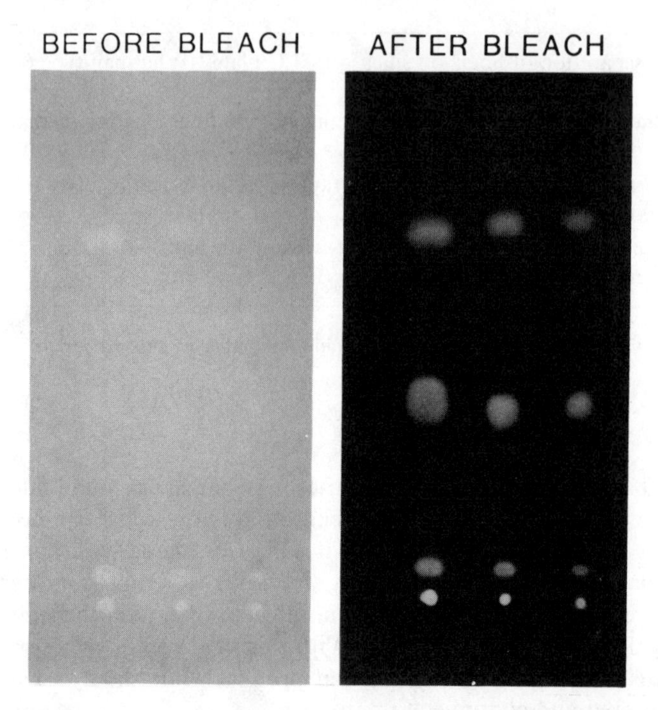

Figure 2.3. Appearance of nile red-stained TLC plate and its decolorization by bleach. Lipid mixtures were spotted on a TLC plate at three concentrations (80, 20, and 5 μg of each lipid type) and chromatographed in hexane–diethyl ether–acetic acid 90:20:1.5 (v/v/v) to separate the lipids into bands. The plate was then dipped in nile red (8 μg/mL), dried, and photographed under ultraviolet light (left photograph). The same plate was then dipped in bleach solution (1/2,500 dilution of Clorox), dried, and photographed again (right photograph). Fluorescence appears white against dark background. Lipid bands in order of increasing mobility to the top are sphingomyelin, cholesterol, trioleoylglycerol, and cholesteryl oleate. Reprinted from Fowler et al. (2) with permission of Lipid Research, Inc.

pids. For amounts of lipid less than 0.6 μg (Figure 2.4, left), the relation between fluorescence and quantity of lipid present is approximately linear. However, with greater lipid amounts, the change in intensity of nile red fluorescence with increasing lipid mass is curvilinear with an upper limit to the amount of nile red florescence that can be achieved. The upper limit of fluorescence obtained varies with lipid class, but the relative height of the curves is constant from experiment to experiment. Similar curvilinear plots are generated from rhodamine-stained plates (6). We extracted nile red dye from lipid spots after staining TLC plates, and found that the relation of nile red extracted to lipid mass is also curvilinear. Thus, we suggest that diffusion

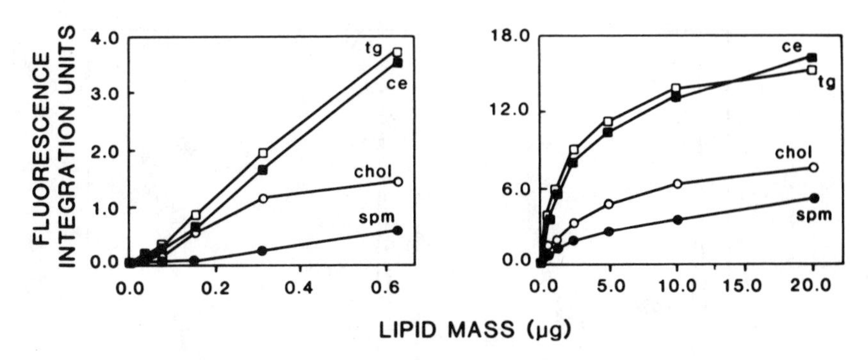

Figure 2.4. Mass calibration plots of nile red-stained lipids. Indicated amounts of cholesterol, cholesteryl oleate, trioleoylglycerol, and sphingomyelin classes were spotted on a single TLC plate. Reprinted from Fowler et al. (2) with permission of Lipid Research, Inc.

of dye into the lipid mass may be rate limiting and produce the curvilinear plots of fluorescence.

The assay, from spotting of samples to quantitating nile red fluorescence, requires about 4 h. The limit of sensitivity for most lipid classes is approximately 100 ng, although as little as 25–30 ng of cholesteryl ester and triacylglycerol is detectable. It is necessary to include a range of lipid calibration standards on each plate for accurate quantitation of an unknown mixture. To study the reproducibility of nile red fluorescence values obtained, we analyzed 1 μg of triacylglycerol 19 times on a single plate and then measured the fluorescence of each spot after staining (2). The standard error of the mean was 1% of the mean fluorescence value. Plate-to-plate variability of nile red fluorescence is also low, with standard error of the mean less than 5% of the mean fluorescence value of the same sample spotted on five separate plates.

An example of the application of the nile red assay is illustrated in Figure 2.5. TLC scan tracings are shown that demonstrate the accumulation of cellular lipids in human skin fibroblasts incubated with low-density lipoprotein for 48 h (2). Cells in a control cell culture incubated in lipoprotein-deficient serum contained no measurable cholesteryl ester, but the triacylglycerol content was 4.5 μg/mg cell protein. In an identical culture incubated with low-density lipoprotein (25 μg protein/mL), cholesteryl ester accumulated to a mass of 23 μg/mg cell protein and the triacylglycerol content increased to 8.6 μg/mg cell protein. No significant increase in the content of unesterified cholesterol was found. Similar analyses of the phospholipid content in these cultures could be carried out after a separate chromatography to resolve those lipids.

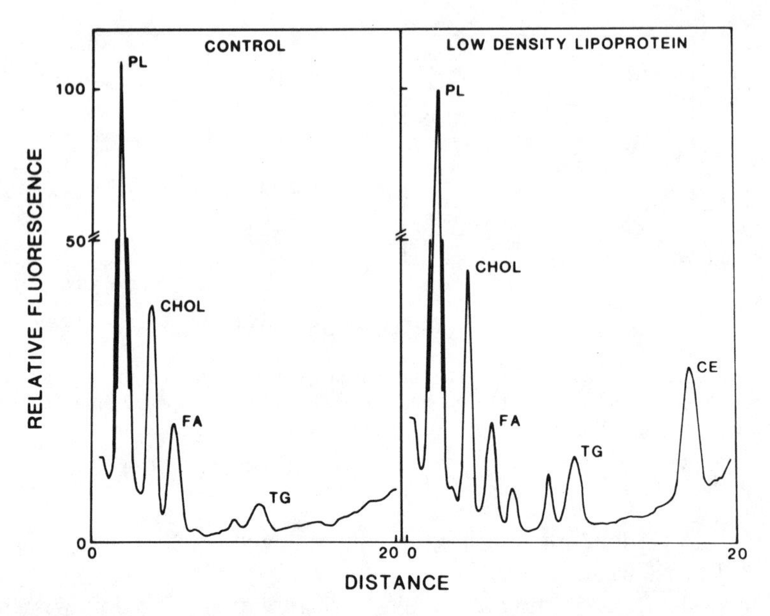

Figure 2.5. Analysis of cellular lipid composition by TLC and nile red staining. Cultured skin fibroblasts were grown and incubated in the presence of lipoprotein-deficient serum alone (control) or in the same medium with the addition of low-density lipoprotein (25 μg of protein/mL). Chloroform extracts, corresponding to 75 μg of cellular protein, were applied to a TLC plate. After chromatographic separation of the lipids in hexane–diethyl ether–acetic acid 146:50:4 (v/v/v), the plate was stained with nile red and decolorized. TLC scan tracings of the fluorescence of the samples are shown. Reprinted from Fowler et al. (2) with permission of Lipid Research, Inc.

TECHNICAL CONSIDERATIONS AND LIMITATIONS

An advantage of the nile red staining procedure is the wide variety of lipid classes that can be detected. However, for purposes of quantitation, the intensity of fluorescene produced with individual lipids is variable (Figure 2.6). Neutral lipids give much stronger nile red fluorescence than other lipid compounds. Thus, for quantitation, calibration standards of each lipid class studied are required. We also found that saturation of the fatty acid moiety of the lipids significantly reduced the intensity of the nile red fluorescence produced (Figure 2.7). Apparently, straight-chain fatty acids are stearically less able to accommodate the presence of neighboring nile red molecules than are the bent chains of unsaturated fatty acids. Thus, for most accurate results, we also suggest use of lipid calibration standards that reflect the fatty acid composition of the samples being quantitated. For example, in the cell

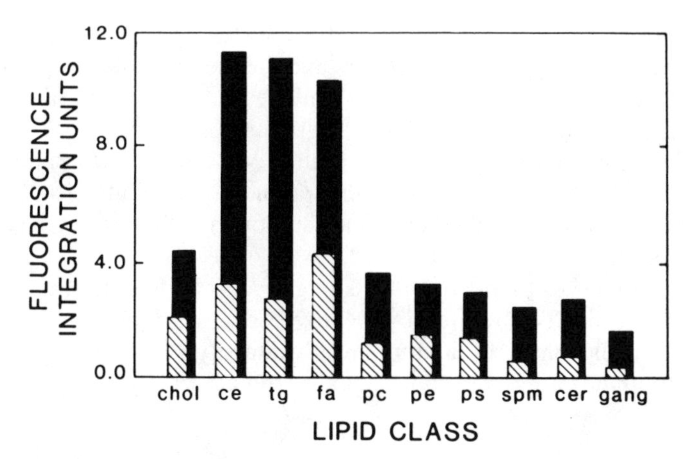

Figure 2.6. Relative fluorescence of various lipid classes stained by nile red. Samples were spotted on a TLC plate in two different amounts (0.5 μg, shaded bars; 2.5 μg solid bars). The lipids examined were respectively, cholesterol, cholesteryl oleate, trioleoylglycerol, oleic acid, phosphatidylcholine, phosphatidylserine, sphingomyelin, cerebroside, and ganglioside. Reprinted from Fowler et al. (2) with permission of Lipid Research, Inc.

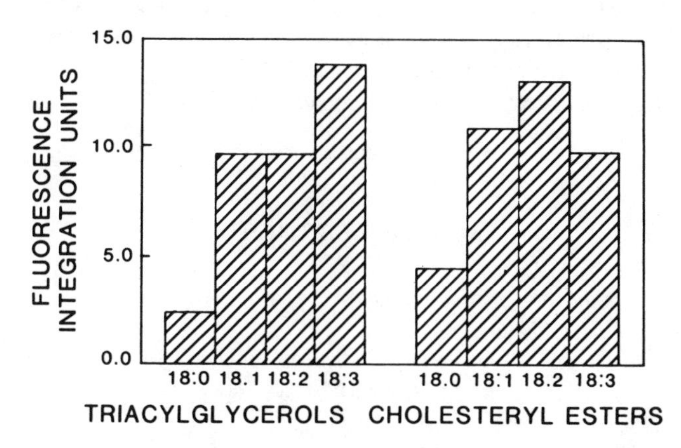

Figure 2.7. Influence of saturation of fatty acid moieties of lipids on the observed nile red fluorescence. Reprinted from Fowler et al. (2) with permssion of Lipid Research, Inc.

culture experiment cited above, we used cholesteryl ester and triacylglycerol standards composed of a 2 : 1 ratio of oleate to stearate fatty acids as esters.

If lipids or other hydrophobic compounds need to be detected for purposes of purification or radioisotopic counting, a quick procedure to reveal the compounds is to spray the chromatographed plate with nile red (40 μg/mL acetone), dry briefly, then mist with distilled water. The plate is then viewed under short-wavelength ultraviolet light and the desired spots scraped off and collected. We found that the amount of nile red extracted from silica gel did not significantly interfere with liquid scintillation counting of radioisotopic labeled lipid samples taken from TLC plates. We view the nile red procedure as non destructive, with the possible exception that the dilute bleach used in our standard protocol could alter some compounds. If that is a concern, the above procedure should be employed.

Those with access to video cameras might consider quantitating nile red-stained lipids on TLC plates by digitized imaging procedures (7). This technique would have the additional advantage of quantitating lipids or other compounds after two-dimensional chromatography.

SUMMARY

The fluorescent dye nile red is recommended as a useful general purpose reagent for the detection and quantitation of a wide variety of lipid and other hydrophobic compounds separated by TLC. Chromatographed TLC plates are dipped into a nile red solution, dried, then dipped into a dilute solution of bleach to decolorize dye bound to silicic acid. Nile red-stained bands are then viewed under ultraviolet light and quantitated by in situ reflectance fluorometry. The method is rapid and can detect as little as 25 ng of some compounds.

ACKNOWLEDGMENTS

Grant support from the American Heart Association is gratefully acknowledged.

REFERENCES

1. J. Sherma and S. Bennett, *J. Liq. Chromatogr.* 6: 1193 (1983).
2. S. D. Fowler, W. J. Brown, J. Warfel, and P. Greenspan, *J. Lipid Res.* 28: 1225 (1987).
3. P. Greenspan, E. P. Mayer, and S.D. Fowler, *J. Cell Biol.* 100: 965 (1985).
4. P. Greenspan and S. D. Fowler, *J. Lipid Res.* 26: 781 (1985).
5. D. T.Downing, *J. Chromatogr.* 38: 91 (1968).
6. L. A. Roch and S. E. Grossberg, *Anal. Biochem.* 41: 105 (1971).
7. S. Inoué, *Video Microscopy* Plenum, New York, (1986).

CHAPTER

3

USE OF BONDED PHASES IN PLANAR CHROMATOGRAPHY

MARTIN GOULD AND THOMAS ENZWEILER

Since the start of thin-layer chromatography (TLC) this analytical method has been subject to continuous development. Standardization of the sorbents used was a necessary prerequisite for the preparation of bulk sorbents and precoated TLC layers with reproducible separation characteristics. For this purpose it was necessary initially to clarify the relations between the physical and chemical properties of the sorbents on the one hand and their chromatographic characteristics on the other hand. Once the relations between primary and secondary parameters of sorbent layers and their chromatographic properties were known, this led to further development of the sorbents and TLC precoated products available at that time, giving rise to high-performance thin-layer chromatography (HPTLC)

Up to now porous silica gel has proved to be the most versatile sorbent and, thus, the sorbent most frequently used in thin-layer chromatography. But in column liquid chromatography and especially in the cases of pharmaceutical and biological analysis, surface-modified column packings have become increasingly important. The most important sorbents in this area are the reversed-phase materials, that is, silica gels with chemically modified, nonpolar, and hydrophobic surfaces. After some time it has also proved possible to produce precoated plates using such surface-modified, reversed-phase sorbents in a coating quality suitable for high-performance thin-layer chromatography.

The surface reactions for the modification of the reversed-phase sorbents in thin-layer chromatography also take place with specific reactive silanes at the accessible silanol groups of the silica gel matrix. New siloxane groupings are formed with the elimination of the silanol groups. At the new siloxane groups aliphatic hydro-carbons are chemically bonded via direct silicon–carbon bonds. The HPTLC precoated plates prepared in this way are named HPTLC RP-2, RP-8, and RP-18.

Figure 3.1 summarizes some important characteristics of the HPTLC RP

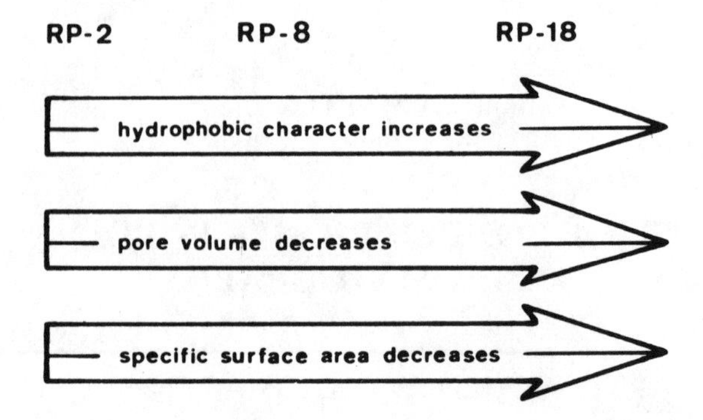

Figure 3.1. Some characteristics of HPTLC RP precoated plates.

plate	: HPTLC pre-coated plate RP-2 F 254s
eluent	: methanol/1N acetic acid 80/20 (v/v)
migration distance	: 7 cm
chamber	: normal chamber without chamber saturation
compounds	: 1 Cholesterol 2 7-Hydroxycholesterol 3 Lithocholic acid 4 Cholic acid methyl ester 5 Cholic acid 6 Dehydrocholic acid
application volume	: 300 nl
detection	: spray reagent: $MnCl_2$ - sulfuric acid heating up to 120° C for 5 min. UV 366 nm

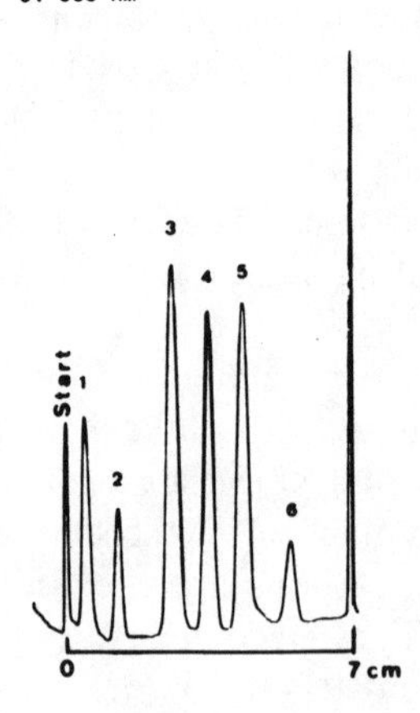

Figure 3.2. Separation of cholesterol and its bile acid metabolites.

pre-coated plates with different alkyl chain lengths. In the sequence from RP-2 to RP-18 the hydrophobic character of the plates increases and the pore volumes and specific surface areas of the modified sorbents decrease in sequence. An application of the HPTLC precoated plate RP-2 is shown in Figure 3.2. Here cholesterol and its bile acid metabolites are separated with the mobile phase of methanol–acetic acid (80:20). At this typical reversed-phase separation the R_f values of the substances increase with increasing polarities.

Figure 3.3 shows the separation of three stilbestrol derivatives with HPTLC RP-8 F 254. In this case the retention of the substances again decreases in accordance with their increasing polarities from the dimethyl ether to the monomethyl ether and finally to the diethylstilbestrol. (1)

A separation with HPTLC RP-18 is shown in Figure 3.4. The steroids methyltestosterone, Reichstein's substance S, and hydrocortisone are separated according to their polarity and hydrophobicity.

plate	: HPTLC pre-coated plate RP-8 F 254s
eluent	: methanol/water 80/20 (v/v)
migration distance	: 5 cm
chamber	: normal chamber without chamber saturation
compounds	: 1 Diethylstilbestrol-dimethyl ether 2 Diethylstilbestrol-monomethyl ether 3 Diethylstilbestrol
application volume	: 200 nl
detection	: in-situ evaluation with TLC/HPTLC scanner (Camag) UV 254 nm

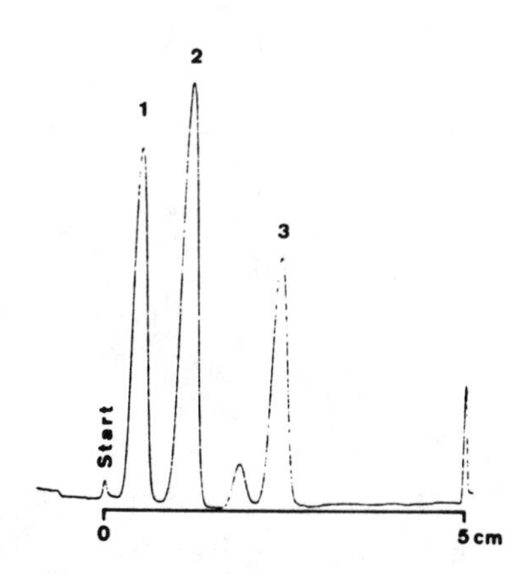

Figure 3.3. Separation of stilbestrol derivatives.

The three HPTLC RP layers mentioned have a high degree of modification and, hence, a very marked hydrophobic character. The extent to which these layers can be wet with aqueous solvent mixtures is limited. At high water contents in the eluent, the capillary forces within the layer are no longer able to counter-act the hydrophobic repulsive force and the solvent does not rise inside the layer. To eliminate these restrictions of the HPTLC RP plates on the choice and composition of solvents and also to enable shorter migration times, especially for highly polar solvents, TLC and HPTLC reversed-phase layers have been introduced that have a reproducible lower degree of modification and in accordance with this a lower degree of hydrophobicity compared with the HPTLC RP plates mentioned above. The plates to be discussed here are the TLC RP-8, the TLC RP-18, and the HPTLC RP-18 W (W means wettable by water) precoated plates. These three RP precoated layers permit even water to be used as solvent for the chromatographic development of the plates.

Another difference between TLC and HPTLC RP layers is the average particle diameter and the particle size distribution of the sorbents used.

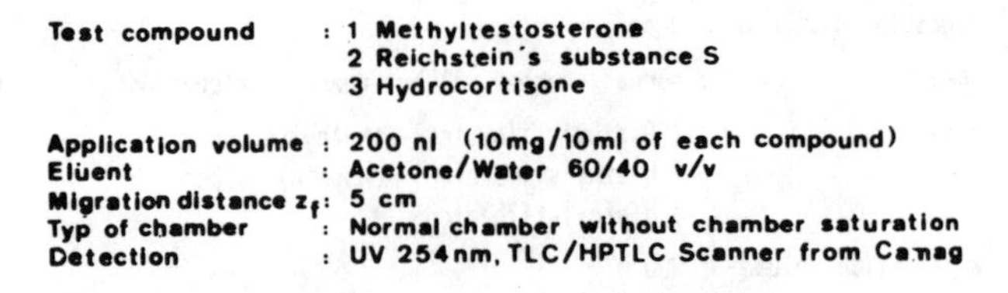

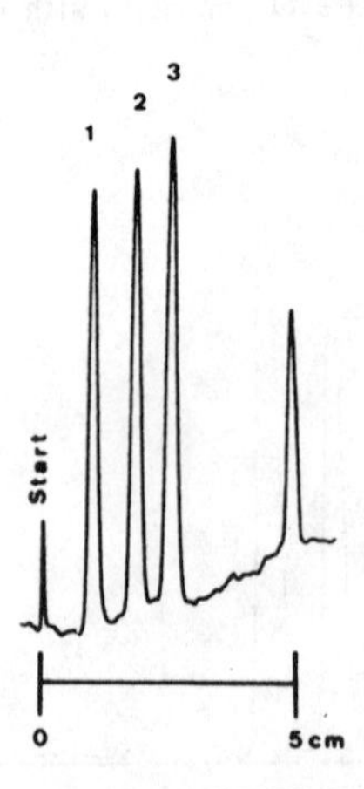

Figure 3.4. Separation of some steroids on HPTLC precoated plate RP-18 F 254s.

Reversed-phase materials for HPTLC layers have a considerably narrower particle size distribution and a smaller mean particle diameter than the corresponding TLC RP types. Consequently, the HPTLC layers have denser and more homogeneous packing and, therefore, a smoother surface structure. This fact offers several advantages in qualitative and in direct quantitat-

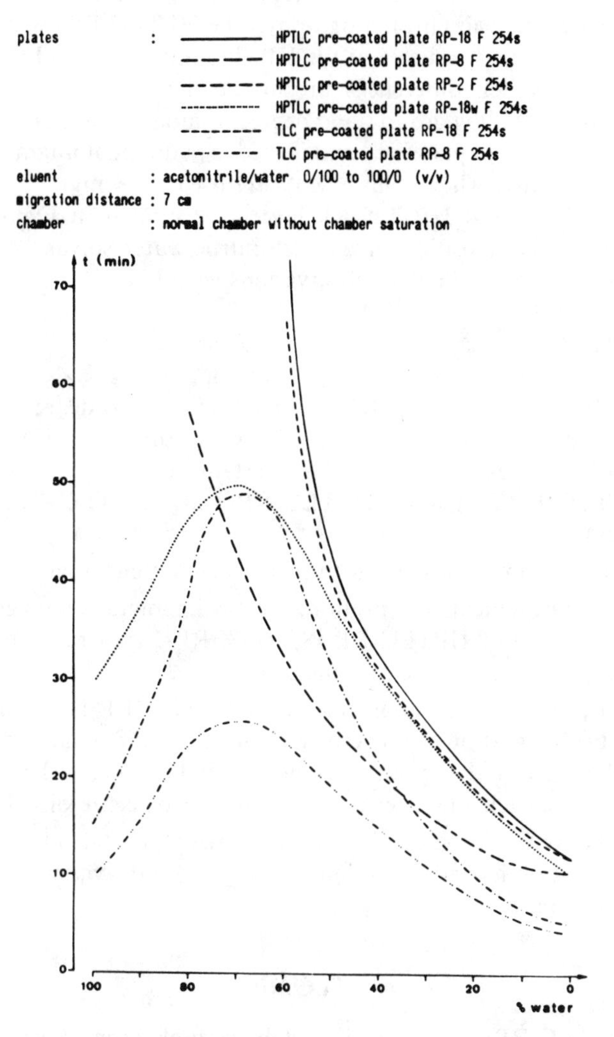

Figure 3.5 Dependence of the migration times of different TLC and HPTLC reversed phase precoated plates on the water content of the eluent. See text.

ive in situ analysis. Furthermore, a special manufacturing method purifies the HPTLC RP layers and eliminates zones of impurities which can appear near the solvent front in the case of TLC RP layers. The coarser particle diameter and, thus, the rougher surface of the TLC RP layers mean that in situ analysis and separation efficiency are lower compared with that of the HPTLC RP plates. Because of their unlimited wettability, however, the TLC RP layers are particularly suitable for use with water-rich eluents and reversed-phase, ion-pair chromatography. The HPTLC RP-18 W precoated plate combines the advantages of the HPTLC RP and the TLC RP layers.

One way in which differences in particle diameters, chain lengths of chemically bonded alkyl groups, and degree of modification of the TLC and HPTLC RP precoated plates are manifested is in different migration characteristics, particularly when polar eluents are used. The migration times t on the different TLC and HPTLC RP layers are plotted in Figure 3.5 as a function of the composition of an acetonitrile/water solvent system. This diagram leads to the following observations:

HPTLC RP

1. All HPTLC RP plates with the maximum possible degree of modification have a limited wettability by water/organic solvent mixtures; therefore, these curves end at certain maximum water contents (HPTLC RP-2 at 80%, HPTLC RP-8 and HPTLC RP-18 at 60% water).
2. Migration times decrease with increasing content of acetonitrile.
3. At identical eluent composition, decreasing migration times appear in the sequence of HPTLC RP-18, RP-8, RP-2 as a result of increased wettability in the same sequence.
4. With pure acetonitrile as eluent, all HPTLC RP layers have nearly identical migration times because there are no differences in wettability in this case and the sorbents of all HPTLC plates have the same average particle diameter and the same particle size distribution.
5. The HPTLC precoated plate RP-18 W with its defined lower degree of modification has no limitation with regard to the water content in the mobile phase.

TLC RP

1. Both TLC RP plates are completely wettable even by water.
2. Each curve of the TLC RP layers (and also the HPTLC RP-18 W) shows a pronounced maximum at an acetonitrile content of 30 to 40%.

3. At identical eluent composition, the migration time of TLC RP-8 is shorter than that of TLC RP-18. The explanation for this is the same as for HPTLC RP.
4. With pure acetonitrile as eluent the two TLC RP plates have an identical migration time, which is shorter than that of the HPTLC RP plates.

A comparison of HPTLC RP layers with the maximum possible degree of modification and TLC RP plates shows that with alkyl groups of identical chain length and with identical eluent composition, the TLC RP plates always have shorter migration times than the HPTLC RP plates. This is due to the larger particle diameters and the less pronounced hydrophobic character of the TLC RP sorbents.

The different degrees of modification of the TLC and HPTLC RP layers, already observed in the different migration characteristics, can also be demonstrated as variations in residual adsorption characteristics. Figure 3.6

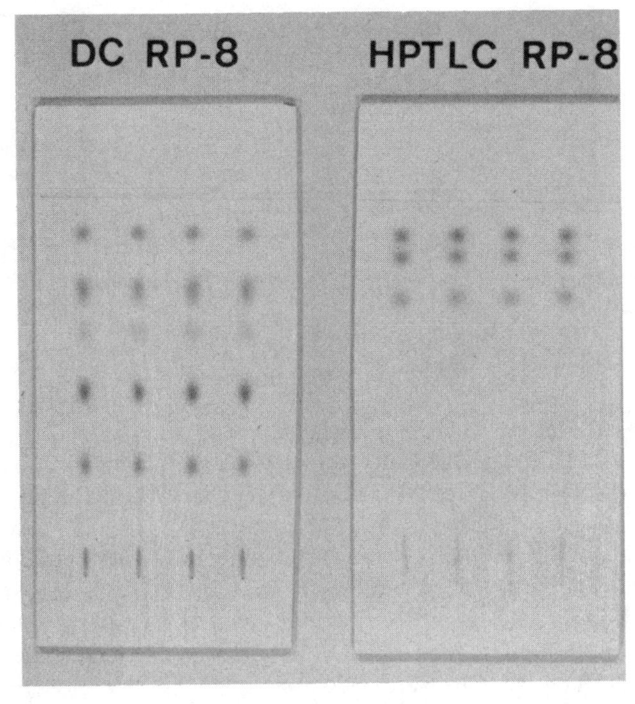

Figure 3.6. Results of separation of a dye mixture showing the difference between C-8 and C-8 on HPTLC.

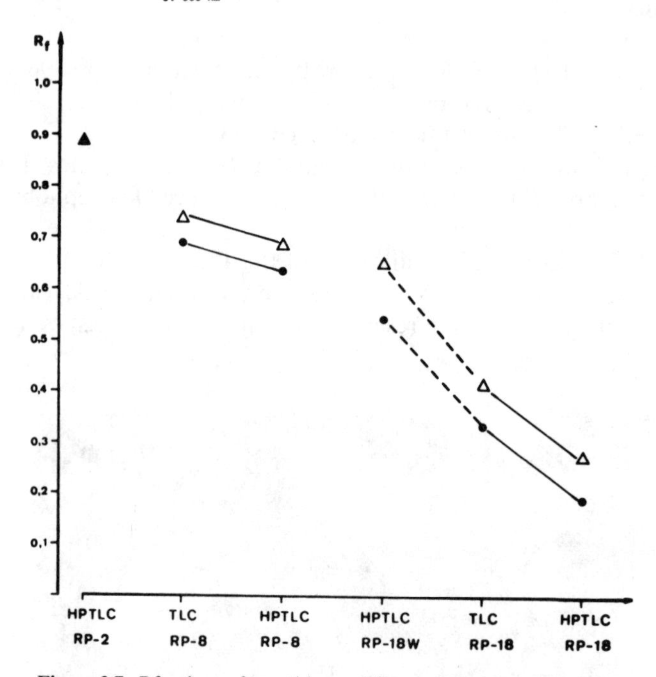

Figure 3.7. Rf-values of steroids on different RP precoated layers.

shows the separation by adsorption chromatography of a lipophilic dye mixture on TLC and HPTLC RP-8 plates, using toluene as solvent. As a result of the lower degree of modification the TLC RP-8 layer shows markedly lower R_f values than the corresponding HPTLC RP-8 layer, owing to more intense residual adsorption of the solute molecules on the unreacted silanol groups. Using the R_f values of the two steroids ergosterol and cholesterol. Figure 3.7 shows the different interactions and the retentions of varying strength on TLC and HPTLC RP precoated layers in the reversed-phase mode using acetone–water (95: 5) as eluent.

The different R_f values on the TLC and HPTLC RP plates can be explained as follows:

1. Because the hydrophobic character of RP plates increases with increasing chain length of the alkyl groups, the R_f values decrease in the following sequences: (a) HPTLC—RP-2, RP-8, RP-18 and (b) TLC—RP-8, RP-18.
2. At identical chain lengths the hydrophobic character of HPTLC RP plates (with the exception of the HPTLC RP-18 W) is more pronounced than that of TLC RP plates, and, therefore, the R_f values on the HPTLC RP plates are lower than those on the corresponding TLC RP plates.
3. R_f values on HPTLC RP-18 W plates are higher than those on TLC RP-18 and HPTLC RP-18 layers because of the lower degree of modification in the case of HPTLC RP-18 W. Therefore, the interaction of the solute molecules based on a reversed-phase retention mechanism is weaker on the HPTLC RP-18 W plate than on the TLC RP-18 and even more pronounced on the HPTLC RP-18 layer.

plate	: TLC pre-coated plate RP-8 F 254s
eluent	: acetone/water 40/60 (v/v)
migration distance	: 7 cm
chamber	: normal chamber without chamber saturation
compounds	: 1 4-Hydroxybenzoic acid propyl ester 2 4-Hydroxybenzoic acid ethyl ester 3 4-Hydroxybenzoic acid methyl ester 4 4-Hydroxybenzoic acid
application volume	: 300 nl
detection	: in-situ evaluation with TLC/HPTLC scanner (Camag) UV 254 nm

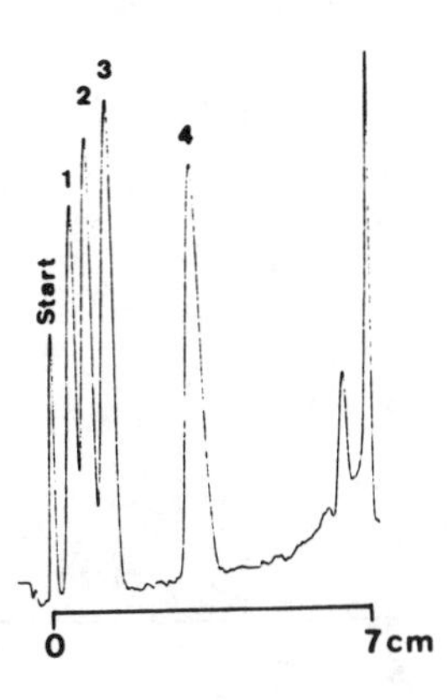

Figure 3.8. Separation 4-hydroxybenzoic acid esters on reversed phase layers.

An example of a separation on a TLC RP-8 layer is shown in Figure 3.8. In this case the preservatives 4-hydroxy benzoic acid and some of its esters were separated using an eluent with a relatively high water content (60%) in acetone. Figure 3.9 shows the chromatogram of the barbiturates revonal, thiogenal, prominal, and luminal on the TLC RP-8 precoated plate with mobile-phase methanol–water (80:20). This example shows that TLC RP plates have good results with only small amounts of water in the mobile phase. An example for the application of the TLC RP-8 precoated plate in ion-pair chromatography is given in Figure 3.10. The nitrogen bases eupaverine chloride, papaverine, codeine and caffeine are separated with a mobile phase of acetone and water with the addition of octane-1 sodium sulfonate as the ion-pair reagent. (2)

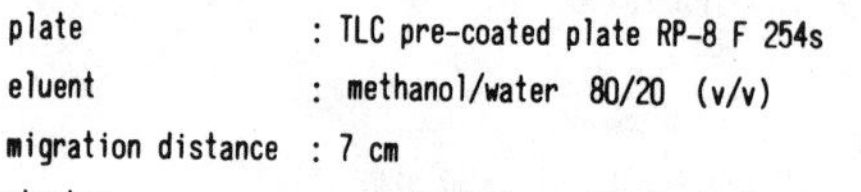

plate	: TLC pre-coated plate RP-8 F 254s
eluent	: methanol/water 80/20 (v/v)
migration distance	: 7 cm
chamber	: normal chamber without chamber saturation
compounds	: 1 Revonal 2 Thiogenal 3 Prominal 4 Luminal
application volume	: 300 nl
detection	: in-situ evaluation with TLC/HPTLC scanner (Camag) UV 254 nm

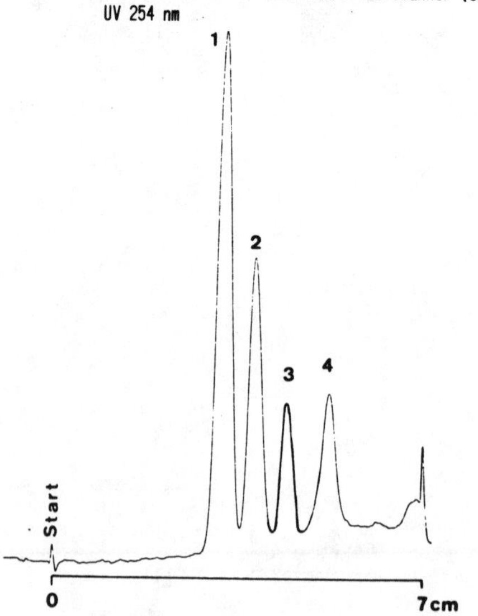

Figure 3.9. Separation of barbiturates on reversed phase TLC.

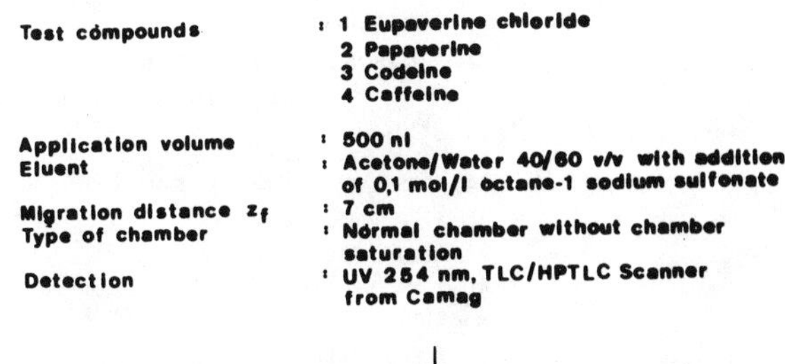

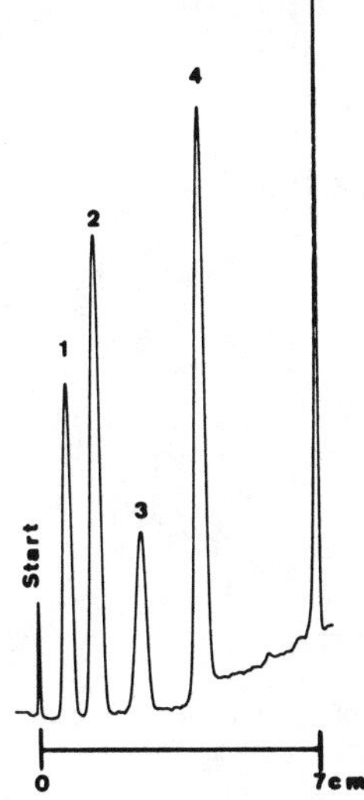

Figure 3.10. Separation of N-compounds on TLC precoated plate RP-8 F254s by ion-pair chromatography.

Figure 3.11 demonstrates separations on a TLC RP-18 layer. The chromatogram of alanine oligomers is an example for the separation of peptides according to the number of segments with the aid of reversed-phase, thin-layer chromatography. The addition of lithium chloride to the mobile phase leads to a more compact spot formation. It is also possible to separate peptides be reversed-phase TLC according to the nature of the segments. This is seen in Figure 3.12 where some penta-peptides are separated on the TLC RP-18 with the addition of tetramethyl ammonium chloride as ion-pairing reagent to the eluent. An example of the use of 100% water as mobile phase with the addition of a only small amount of lithium chloride on an HPTLC RP-18 W is the separation of nicotinic acid and isonicotinic acid shown in Figure 3.13.

The following can be summarized with regard to reversed-phase precoated thin-layers: To overcome the disadvantages of nonpolar impregnated

plate : TLC pre-coated plate RP-18 F 254s
eluent : acetonitrile/ethanol/water 90/5/10 (v/v) with addition of 0,1 mol/l lithium chloride
migration distance : 12 cm
chamber : normal chamber without chamber saturation
compounds : 1 Ala-Ala-Ala-Ala-Ala
2 Ala-Ala-Ala-Ala
3 Ala-Ala-Ala
4 Ala-Ala
detection : spray reagent : ninhydrin
heating up to 110° for 15 min.

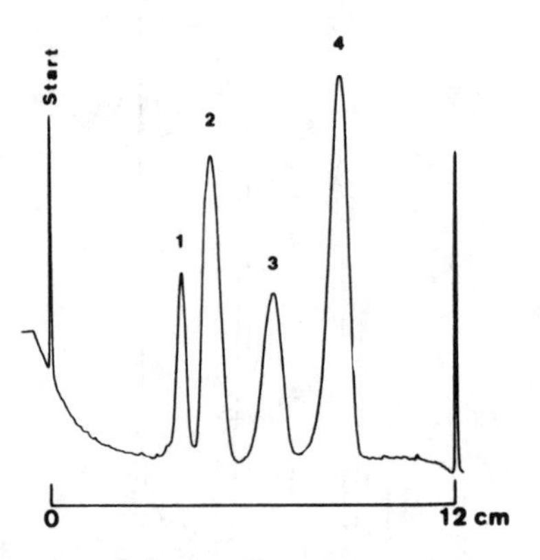

Figure 3.11. Separation of alanine oligomeres on C_{18} layers.

layers, reversed-phase precoated plates have been developed with chemically bonded alkyl groups in two different product lines, the HPTLC RP and the TLC RP plates. In both cases the RP layers were prepared with alkyl groups of different lengths to achieve a wide range of selectivity.

The HPTLC RP layers with the exception of HPTLC RP-18 W, have a maximum degree of modification. Therefore, they show a pronounced hydrophobic character and are limited with regard to the maximum water content in the mobile phase. Because of the extremely narrow particle size distribution and the optimized small mean particle diameter, the HPTLC RP layers show very homogeneous packing and a smooth layer surface, resulting in low baseline noise when they are evaluated by in situ scanning. Furthermore, the HPTLC RP plates show almost no zones of impurities when they are developed with mobile phases.

To overcome the limitation of the restricted water content in the eluent in the case of the HPTLC RP-2, RP-8, and RP-18, TLC RP layers have been introduced with a defined and reproducible lower degree of modification.

plate : TLC pre-coated plate RP-18 F 254s
eluent : acetonitrile/1N hydrochloric acid 90/10 (v/v)
with addition of 0.1 mol/l tetramethylammonium chloride
migration distance : 8 cm
chamber : normal chamber without chamber saturation
compounds : 1 Glu-Thr-Tyr-Ser-Lys
2 Ala-Ala-Ala-Ala-Ala
3 Gly-Gly-Asn-Asp-Ala
4 Ala-Ala-Tyr-Ala-Ala
5 Leu-Trp-Mel-Arg-Phe (all 0.2 %)
application volume : 700 nl
detection : spray reagent: ninhydrin
heating up to 120° C for 10 min.

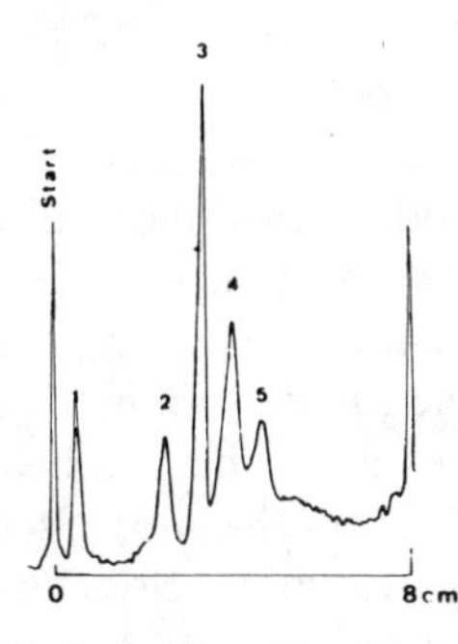

Figure 3.12. Separation of pentapeptides on C_{18} layers.

plate : HPTLC pre-coated plate RP-18w F 254s
eluent : water, with addition of 0,1 mol/l LiCl
migration distance : 5 cm
chamber : normal chamber without chamber saturation
compounds : 1 Nicotinic acid (0,2%)
2 iso-Nicotinic acid (0,5%)
application volume : 200 nl
detection : in-situ evaluation with TLC/HPTLC scanner (Camag)
UV 270 nm

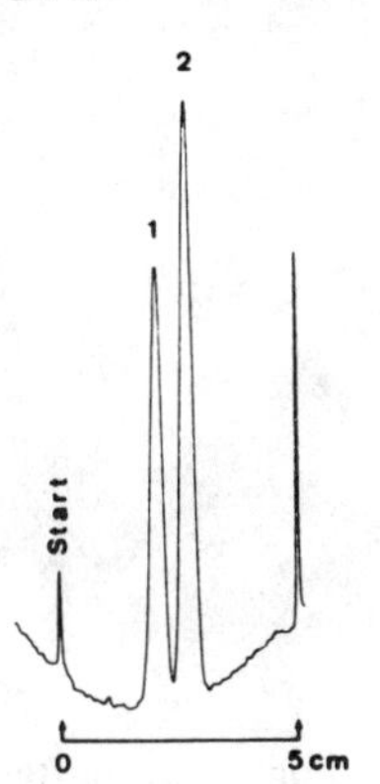

Figure 3.13. Separation of nicotinic acid and isonicotinic acid on HPTLC reversed phase.

These TLC RP are completely wettable and therefore can be used with water as mobile phase.

To combine the advantages of HPTLC RP and TLC RP layers the HPTLC RP-18 W have been developed. Precoated layers with nonmodified silica gel and HPTLC RP-18 have been modified with a concentrating zone. This makes it possible to use the complete separation efficiency of the HPTLC plates even when larger sample volumes are applied. Furthermore, the plates with a concentrating zone allow a very easy sample application and in some cases they can render special sample preparations superfluous.

The concentrating zone has a dimension of 25 mm in the direction of chromatography and covers the whole lower edge of the plate. The sorbent of the concentrating zone is in this case a silanized Silica 50000. Silica 50000 is synthetically produced, inert, and porous silicon dioxide. The concentrating zone and the chromatography separation layer border on each other in a sharp boundary line. The different concentrating and separation steps on an HPTLC with concentrating zone can be seen in Figure 3.14. The mixture of dyestuffs used here can be applied at any desired height on the concentrating zone without having any influence on the subsequent position of the separated substance spots. The concentrating process can be observed in a total of

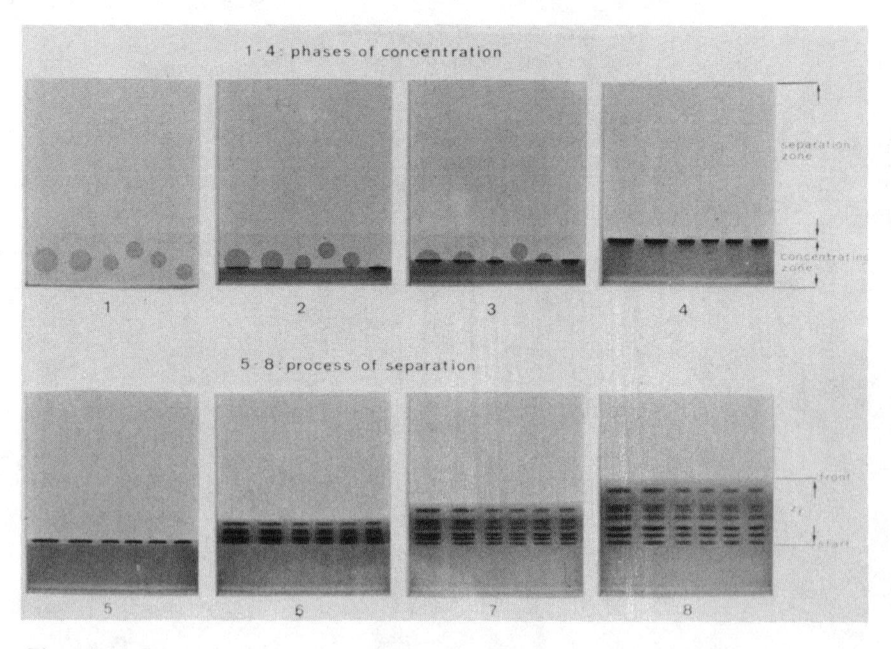

Figure 3.14. Progression in the separation of a dye mixture using the preconcentrating zone. See text.

eight phases photographed one after the other starting from the top left and moving to the right. The phase of highest concentration is obtained in the upper row right and in the row below on the left side. This is specifically the time at which the solvent, together with the bands of the spots applied, which have been pushed together, have reached the boundary line between the two layer sections. Actual chromatography starts from this boundary line and can be followed from left to right in the lower row. Furthermore, this figure shows that after application of the substance mixture as a circular spot, concentration to a narrow start line takes place automatically during the development process. This concentrating phase passes into chromatographic separation without interruption and without a change of the mobile phase. Concentration of the spots in the direction of chromatography increases the separation efficiency of this chromatographic system, especially when larger sample amounts are used.

Figure 3.15 shows the separation of different fats and oils on an HPTLC RP-18 with concentrating zone. Each of these fats and oils has a specific

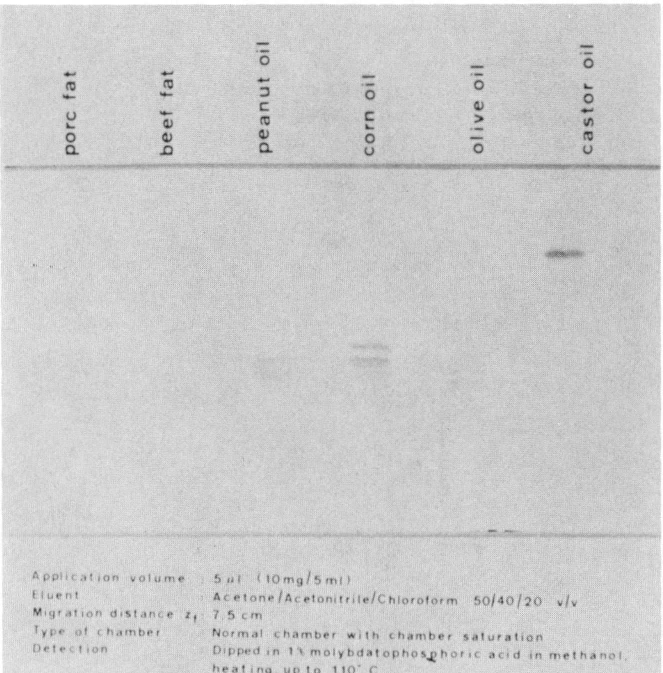

Figure 3.15. Separation of fats and oils on HPTLC precoated plate RP-18 with concentrating zone (according to N. Schmidt, Offenbach).

elution pattern and therefore this separation can be used to identify the sources of fats for example, in foodstuffs. Figure 3.16 demonstrates the separation of some chlorophenols on an HPTLC RP-18 layer with concentrating zone. Another important group of substances that can be separated, preferably on HPTLC RP-18 with concentrating zone, are polycyclic aromatic hydrocarbons. Figure 3.17 demonstrates the separation of five different aromatic hydrocarbons at a temperature of −18 °C. These substances are relevant for the analysis of water and the environment. The low temperature led to a decrease of spot diffusion, because of an increase of separation efficiency.

The hydrophobic, nonpolar modified sorbents and recently polar bonded, hydrophilic stationary phases are often used in liquid column chromatography for separation of hydrophilic or charged substances. The important functional groups on the surface of the sorbents in this field are ion exchangers and diol, cyano, and amino modifications.

plate	: HPTLC pre-coated plate RP-18 F 254s with concentrating zone
eluent	: methanol/water 90/10 (v/v)
migration distance	: 7 cm
chamber	: normal chamber without chamber saturation
compounds	: 1 Pentachlorophenol 2 2.4,6-Trichlorophenol 3 2.4-Dichlorophenol 4 4-Chlorophenol (all 0,1%)
application volume	: 2 µl
detection	: in-situ evaluation with TLC/HPTLC scanner (Camag) UV 254 nm

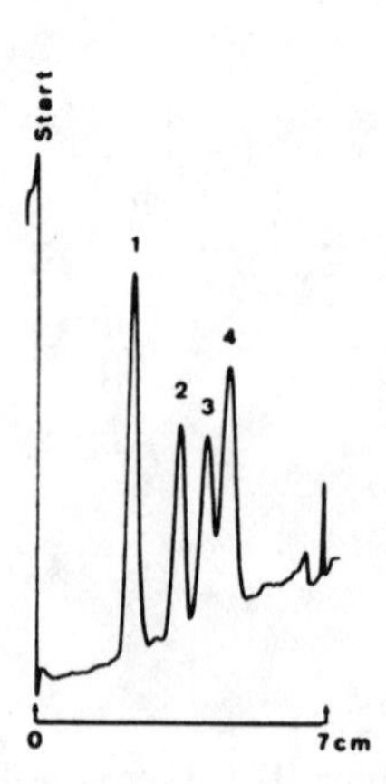

Figure 3.16. Separation of chlorophenols an HPTLC C_{18} layers.

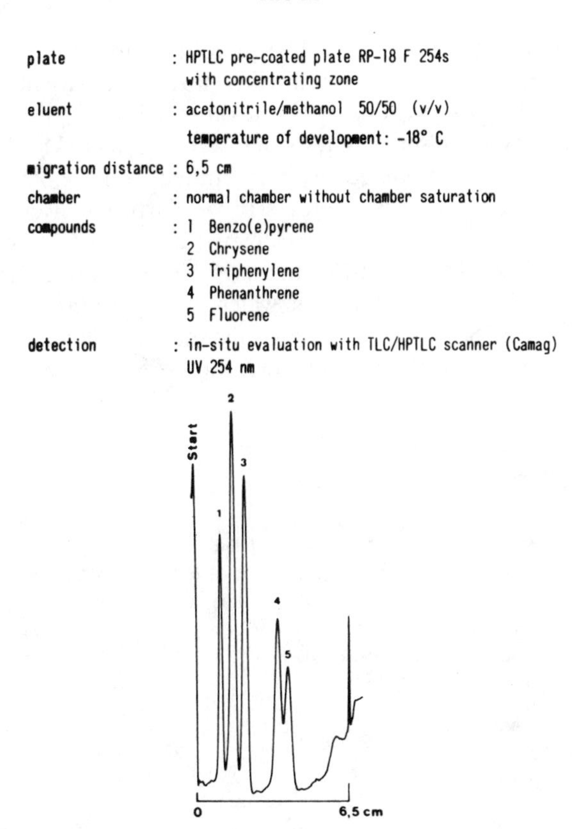

Figure 3.17. Separation of some polynuclear aromatic hydrocarbons relevant for analysis of water and environment.

In analogy to the hydrophilic modified stationary phases used in liquid column chromatography, an HPTLC precoated layer has been developed in which amino groups are bonded chemically to a silica gel matrix. This HPTLC layer (NH_2 F 254s) is based on the same silica gel type 60 as used for the HPTLC silica gel 60 and the modified HPTLC RP-2, RP-8, RP-18, and RP-18 W. The layer also incorporates an acid-stable fluorescent indicator, which responds at 254 nm with a light blue fluorescence. The amino group takes the form of a short-chain alkylamino function bonded chemically to the surface of the silica gel. The chromatographic properties of the amino layers are largely governed by the polar, basic nature of the amino group. This is the reason the HPTLC NH_2 plate can be regarded as a weak base anion exchanger. Various additional surface properties result, however, from the nature of this modification. The alkylamino groups bonded to the silica gel surface may enter into weak hydrophobic interactions with sample substances of appropriate structure. Furthermore, selectivity may be affected by nonreacted silanol groups of the silica gel matrix.

Figure 3.18 shows two series of tests that were conducted to determine the polarity of the amino modified layer in comparison to other stationary phases. The retention of three steroids (hydrocortisone, Reichstein's substance S, methyltestosterone) was investigated on HPTLC precoated plates silica gel 60, RP-2, RP-8, RP-18, and NH_2, using two different solvent systems on each plate, namely acetone/water (Figure 3.18*a*) and chloroform/methanol (Figure 3.18*b*). These two diagrams show the dependence of the R_f values of the steroids on the type of the layer used.

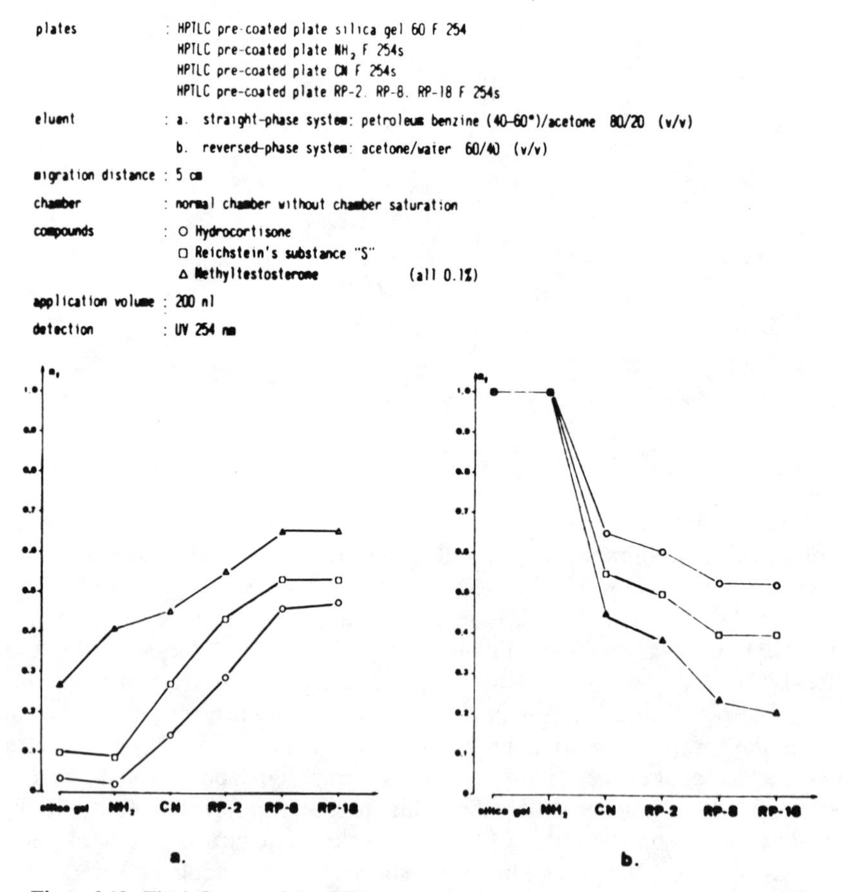

Figure 3.18. The influence of the different stationary phases on the Rf-values of steroids

With the polar, aqueous eluent acetone/water the substances are not separated on the silica gel 60 plate or on the amino modified plate. In both cases they are eluted with the solvent front. Separation occurs, however, on the reversed-phase layers. The R_f values of the steroids decrease in this case as the chain lengths of the RP layers increase. The elution sequence on the RP plates arranged in order of increasing R_f values is methyl testosterone, Reichstein's substance S, and hydrocortisone, which has the weakest retention. The opposite occurs when the nonpolar eluent chloroform/methanol is used. Then there is no retention on the RP plates and all three steroids are eluted here with the solvent front. On the two polar plates, on the other hand, there are marked differences in the retention. On the amino plate the R_f values of the steroids are somewhat larger than on the silica gel 60 plate, although no decisive difference with regard to selectivity is noticeable. On both polar layers hydrocortisone has the smallest R_f value, followed by Reichstein's substance S and methyltestosterone, which has the highest R_f value. The sequence of the substances is thus the exact opposite of that using

plate	: HPTLC pre-coated plate NH_2 F 254s
eluent	: acetone/water 30/70 (v/v) with addition of 0,2 mol/l NaCl
migration distance	: 5 cm
chamber	: normal chamber without chamber saturation
compounds	: 1 UTP 2 UDP 3 UMP 4 UDP-Glucose 5 Uridine
detection	: in-situ evaluation with TLC/HPTLC scanner (Camag) UV 254 nm

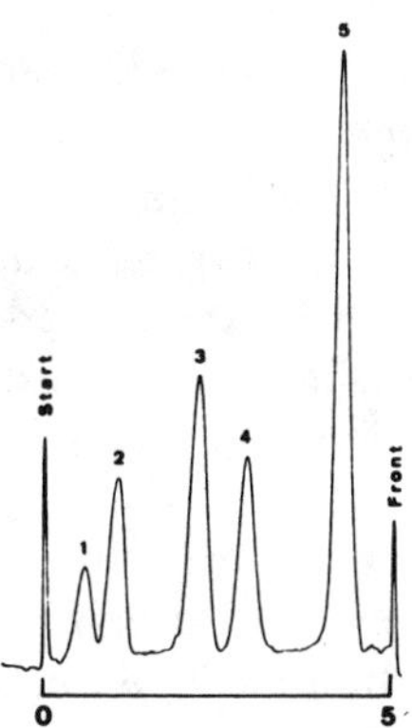

Figure 3.19. Separation of uridine and its phosphates on amine modified layers.

reversed phase with the polar, water-containing eluent. As the two diagrams show, the polarity of the amino plate is lower than that of the silica gel 60 but its lipophilic character is less pronounced than that of the HPTLC RP. Thanks to its hydrophilic modification, the amino layer can be wet with water without addition of organic solvents or salts. Development of the chromatogram with water or with mobile phases of high water content is unproblematic. The possibility of working with pure organic solvents or with mixtures of any composition is also available. As already described, the surface alkylamino groups mainly determine the chromatographic properties of the HPTLC NH_2 layers. Hence, it is to be expected that there will be a special polyanion selectivity in aqueous solvents.

Figure 3.19 shows the chromatographic separation of some uridine nucleotides on an HPTLC NH_2 precoated plate. the addition of sodium chloride to the eluent in this case suppresses secondary interactions and the consequent spreading effect of the spots. The nucleotides are retained in a degree determined by their charges. The highest retention in this case is obtained for uridine–triphosphate with a total charge number of -4 and the uncharged compound uridine shows the lowest retention, that is to say, the highest R_f value. A separation of nucleotides with the same number of charge but with different molecular rates and polarities on an HPTLC NH_2 plate is shown in Figure 3.20. The test compounds are adenosine monophosphate, guanosine monophosphate, and uridine monophosphate.

Figure 3.21 demonstrates the separation of some derivatives of purine on the HPTLC precoated plate NH_2. (3) The purines investigated differ in the number of their hydroxyl and amino groups. In this case the R_f values increase in the sequence uric acid, xanthine, and hypoxanthine with decreasing number of hydroxyl groups. Guanine has both a hydroxyl and an amino group, but is retarded less than hypoxanthine. The NH_2 group on the purine ring obviously reduces the interaction of the molecule with the amino of the solvent. Adenine has no hydroxyl group but one NH_2 group. Therefore, it is retarded the least and has the largest R_f value in the series of the purines investigated here. Figure 3.22 shows an example of the application of the HPTLC NH_2 with a water-free organic eluent. (4) The chromatogram demonstrates the separation of xanthine and its *N*-methyl derivatives. The substances here differ only slightly from each other in their molecular weights, but they differ considerably with respect to their polarity. With pure methanol as eluent the R_f values of the purines increase with increasing number of methyl groups. In this case this means that xanthine has the smallest R_f value and caffeine the largest. An interesting aspect of this separation is that the resolution of the isomeric dimethyl derivatives theophylline and theobromine is larger than the resolution between theobromine and caffeine, which differ in one methyl group. Figure 3.23 shows the

Plate: HPTLC NH_2 F 254s
Eluent: ethanol/water = 30/70 (v/v) with addition of
0,18 mol/l NaCl
Migration distance z_f: 5 cm
Normal chamber without chamber saturation
Detection: in-situ evaluation with
TLC/HPTLC scanner (Camag) UV 254 nm

(1) GMP
(2) UMP
(3) AMP

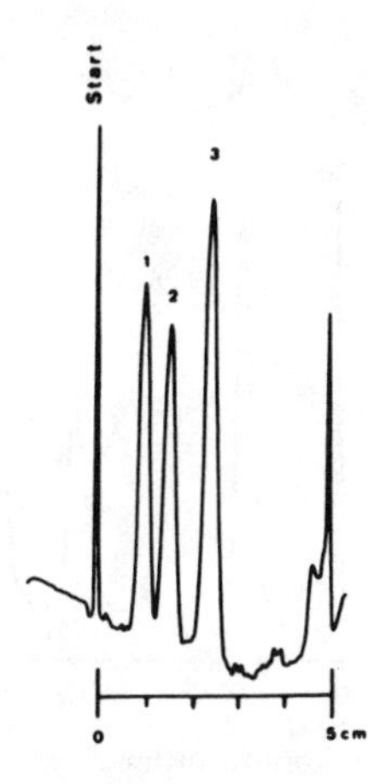

Figure 3.20. Separation of some nucleosidemonophosphates.

Plate: HPTLC NH_2 F 254s
Eluent: ethanol/water = 80/20 (v/v) – saturated with NaCl
Migration distance z_f: 7 cm
Normal chamber without chamber saturation
Detection: in-situ evaluation with
TLC/HPTLC scanner (Camag) UV 270 nm

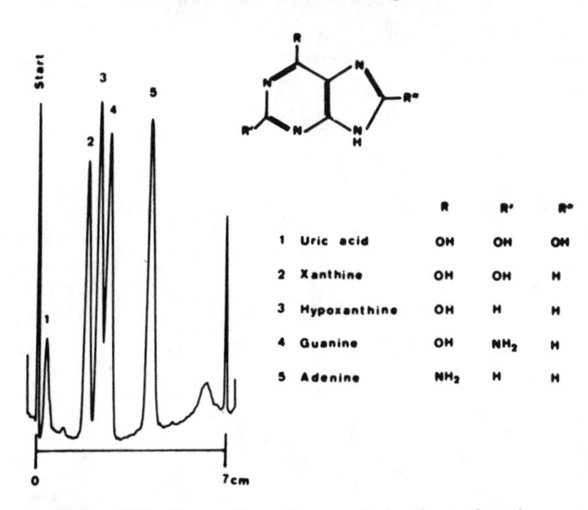

		R	R'	R''
1	Uric acid	OH	OH	OH
2	Xanthine	OH	OH	H
3	Hypoxanthine	OH	H	H
4	Guanine	OH	NH_2	H
5	Adenine	NH_2	H	H

Figure 3.21. Separation of some derivatives of purine.

Plate	: HPTLC pre-coated plate NH_2 F 254s
Eluent	: methanol
Migration distance, z_f	: 5 cm
Detection	: in situ evaluation with a TLC/HPTLC scanner made by Camag (Switzerland), UV 254 nm
Sample substances	: xanthine (1) 0.03%, theophylline (2) 0.03%, theobromine (3) 0.06%, caffeine (4) 0,02%
Sample sizes	: 100 nl

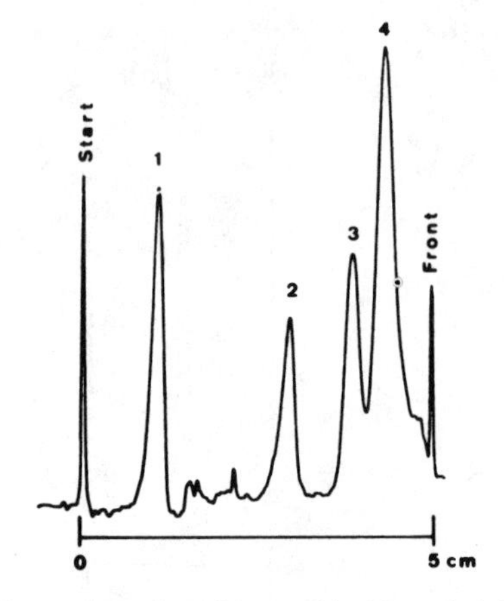

Figure 3.22. Separation of xanthine and its N-methyl derivatives.

plate	: HPTLC pre-coated plate NH_2 F 254s
eluent	: chloroforme/methanol 80/20 (v/v)
migration distance	: 7 cm
chamber	: normal chamber without chamber saturation
compounds	: 1 Tyrocidine 2 Gramicidin (all 0.1%)
application volume	: 300 nl
detection	: in-situ evaluation with TLC/HPTLC scanner (Camag) UV 270 nm

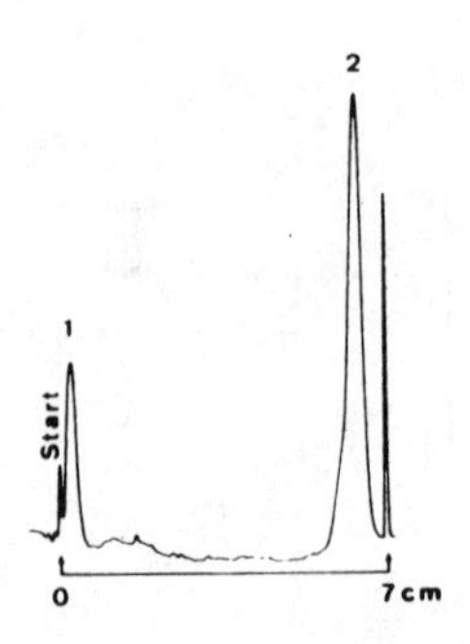

Figure 3.23. Separation of gramicidin and tyrocidine.

separation of two cyclopeptide antibiotics, gramicidin and tyrocidine of the HPTLC NH_2 with a water-free organic mobile phase.

As a further hydrophilic modification in thin-layer chromatography a cyano phase has been developed. The base material taken for the HPTLC CN F 254s is the same silica gel 60 that has been used for all previously mentioned surface-modified HPTLC layers. In terms of mean particle size and particle size distribution, the cyano-modified layer is identical with the modified and non-modified HPTLC layers. The cyano function is chemically bonded to the surface of the silica gel skeleton in the form of a γ-cyanopropyl group.

As a means of categorizing the polarity of the HPTLC CN within the range of modified and nonmodified HPTLC layers, the various retentions for selected steroids in two different solvent systems as shown before in the case of the amino layers were investigated. Figure 3.24 demonstrates the different R_f values for the steroids hydrocortisone, Reichstein's substance S, and methyltestosterone on the HPTLC silica gel 60, NH_2, CN, RP-2, RP-8, and RP-18. In the normal system in Figure 3.24*a* the solvent consisted of petroleum benzine and acetone, while acetone and water were used in the

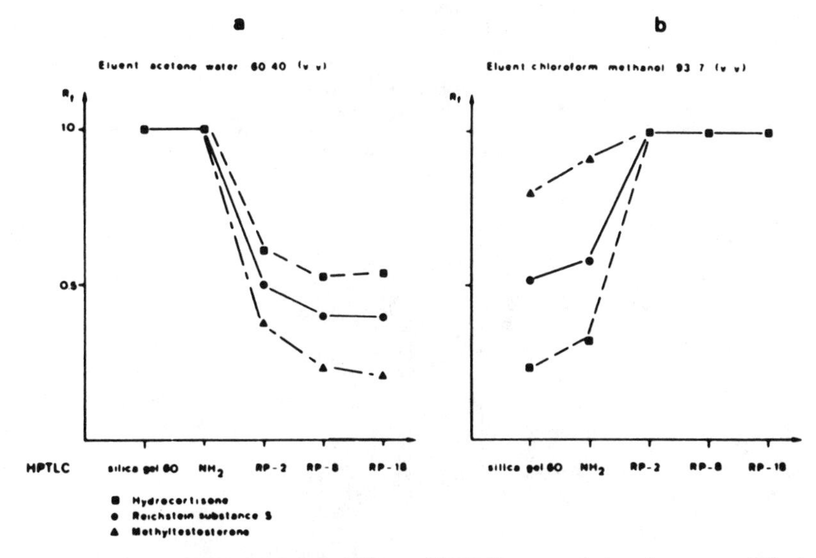

Figure 3.24. A comparison of polarity of different HPTLC precoated plates by means of Rf values of steroids with two different eluents.

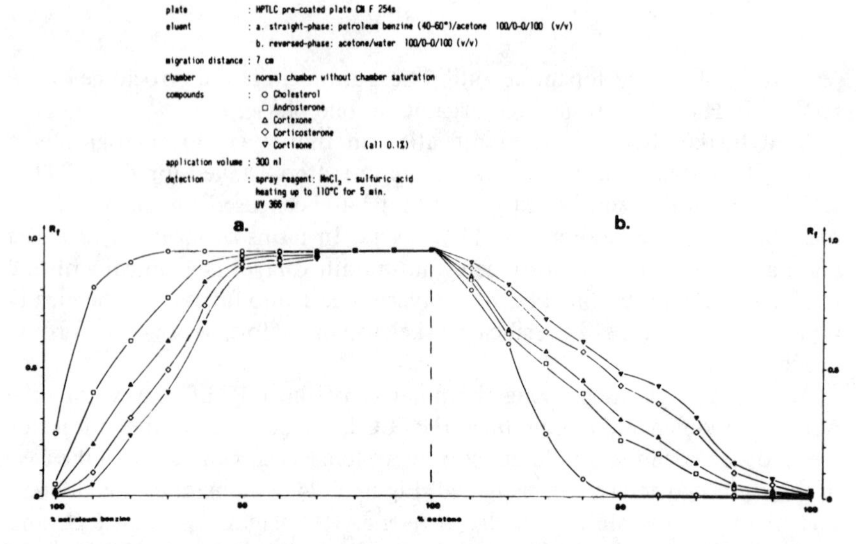

Figure 3.25. The dependence of Rf-values of some steroids on the eluent composition using the HPTLC pre-coated plate of F 254g is a straight-phase and a reversed-phase system.

plate : HPTLC pre-coated plate CN F 254s
eluent : petroleum benzine (40–60°C)/acetone 80/20 (v/v)
migration distance : 7 cm
chamber : normal chamber without chamber saturation
compounds : 1 Estriol
2 Estradiol
3 Estrone (all 0,1%)
application volume : 200 nl
detection : spray reagent: $MnCl_2$ -sulfuric acid
heating up to 120° C for 10 min.
UV 366 nm

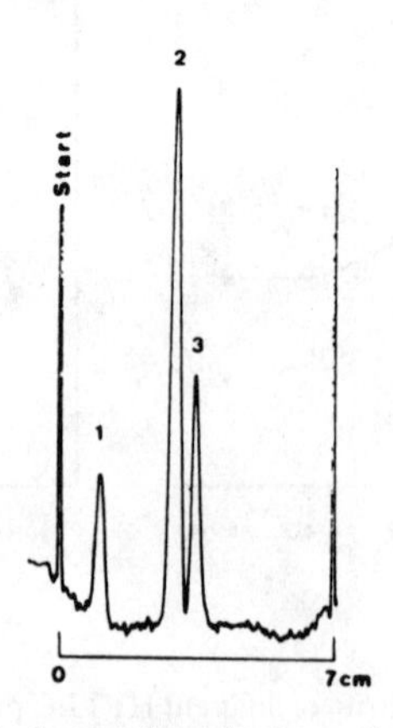

Figure 3.26. Separation of estrone derivatives.

reversed-phase system in Figure 3.24*b*. Both the normal and the reversed-phase chromatographic systems show that, as far as polarity is concerned, the cyano-modified plate is positioned between the NH_2 and the RP-2 sorbents.

Figure 3.25 illustrates the fact that HPTLC CN can be used successfully with mobile phases having completely different polarities and thus also differing retention mechanisms in the case of steroids. When pure petroleum benzine is used (a nonpolar solvent), the steroids have extremely high retentions as a result of pronounced adsorptive interactions. Under these chromatographic conditions the system has a low selectivity. Increasing the polarity by adding acetone to the mobile phase lessens retentions and raises selectivity. Once the eluent has a medium polarity (achieved by adding a large amount of acetone or by use of pure acetone) the steroids without exception are retained only slightly and they have virtually the same R_f values. In other words, this chromatographic system is no longer selective. Raising the polarity of the mobile phase still further by adding water produces a

plate	: HPTLC pre-coated plate CN F 254s
eluent	: petroleum benzine (40-60°)/acetone 80/20 (v/v)
migration distance	: 7 cm
chamber	: normal chamber without chamber saturation
compounds	: 1 Testosterone 2 5-Dehydro-Androsterone 3 Androsterone (all 0,2%)
application volume	: 200 nl
detection	: spray reagent: 5% methanolic perchloric acid heating up to 110°C for 5 min. UV 366 nm

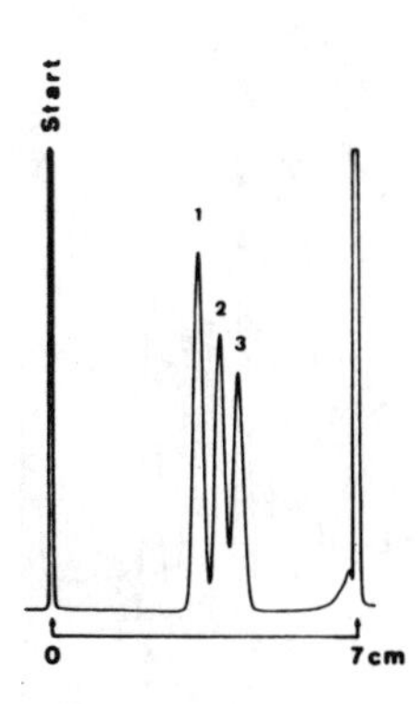

Figure 3.27. Separation of androgens on a cyano modified layer.

reversed-phase system in which both retention and selectivity are raised. Further-more, the retention sequence is diametrically opposed to the normal system already described. Finally, when a maximum-polarity eluent consisting of pure water is used, the selectivity reaches a minimum again because of the consequent very high retention of the steroids.

Figure 3.26 shows the separation of estrogens with petroleum benzine and acetone as normal eluents. Based on the retention mechanism prevailing here, estriol is retained the most because this is the most polar substance. Estradiol and estrone follow with higher R_f values according to their decreasing polarities. Figure 3.27 illustrates the separation of some androgens based on petroleum benzine-acetone, again a normal phase solvent system. (5)

A further use of the HPTLC CN layers is in reversed-phase chromatography in cases in which highly polar aqueous mobile phases are required.

plate	: HPTLC pre-coated plate CN F 254s
eluent	: methanol/1-propanol/water 25/20/55 (v/v/v) with addition of 0,1 mol/1 lithium chloride
migration distance	: 8 cm
chamber	: normal chamber without chamber saturation
compounds	: 1 PTH-DL-Lysine 2 PTH-DL-Proline 3 PTH-DL-Tyrosine 4 PTH-DL-Glutamic acid 5 PTH-DL-Glutamine 6 PTH-DL-Histidine 7 PTH-DL-Cysteic acid (all 0,2%)
application volume	: 200 nl
detection	: in-situ evaluation with TLC/HPTLC scanner (Camag) UV 254 nm

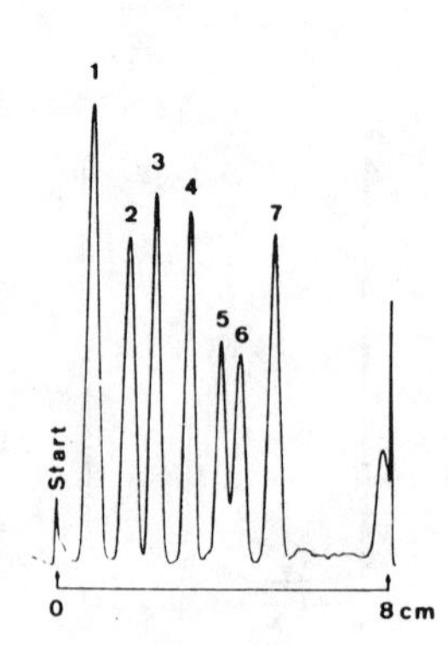

Figure 3.28. Separation of PTH amino acids on a cyano modified layer.

The separation of PTH amino acids is shown in Figure 3.28. The eluent consists of methanol–propanol–water (25:20:55) with the addition of lithium chloride. Lithium chloride in this case reduced tailing. Predictably, in view of the reversed-phase mechanism encountered here, the least polar substance, PTH lysine, is retained most and the most polar substance, in this case PTH cysteic acid, is retained the least.

A further example for the use of a highly aqueous mobile phase in conjunction with HPTLC CN is the separation of various alkaloids as illustrated in Figure 3.29. Here again, an inorganic salt (ammonium bromide) has been added to the mobile phase to reduce tailing. The retention sequence is determined by the polarity of the alkaloids. Emetine chloride is the least polar substance in the test series; consequently, it has the lowest R_f value in this RP system. The other test substances have ascending R_f values in keeping with their increasing hydrophilic character. Figure 3.30 shows

plate	: HPTLC pre-coated plate CN F 254s
eluent	: methanol/1-propanol/water/ammonia solution 25% 30/20/50/1 with addition of 0,1 mol/l ammonium bromide
migration distance	: 7 cm
chamber	: normal chamber without chamber saturation
compounds	: 1 Emetine chloride (0,4%) 2 Quinine (0,1%) 3 Atropine (3,0%) 4 Brucine (0,1%) 5 Scopolamine (3,0%) 6 Caffeine (0,3%) 7 Nicotinamide (0,2%) 8 DL-Tropic acid (3,0%)
application volume	: 300 nl
detection	: in-situ evaluation with TLC/HPTLC scanner (Camag) UV 254 nm

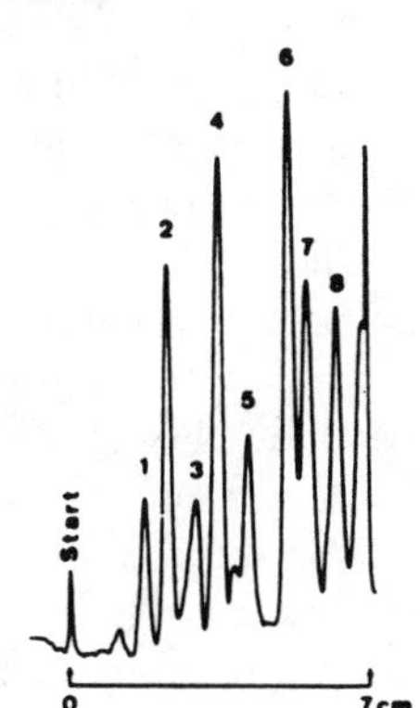

Figure 3.29. Separation of alkaloids.

plate	: HPTLC pre-coated plate CN F 254s
eluent	: ethanol/water 20/80 (v/v) with addition of 0,1 mol/l tetraethylammonium chloride
migration distance	: 7 cm
chamber	: normal chamber without chamber saturation
compounds	: 1 Benzoic acid (2,0%)
	2 Sorbic acid (0,2%)
application volume	: 200 nl
detection	: in-situ evaluation with TLC/HPTLC scanner (Camag) UV 265 nm

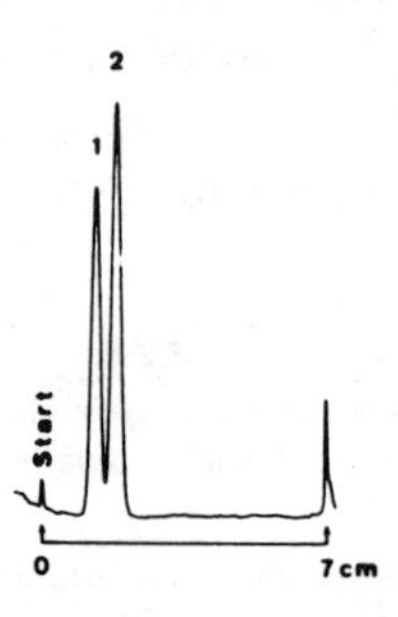

Figure 3.30. Separation of benzoic acid and sorbic acid.

separating on of the two preservatives, benzoic acid and sorbic acid, on a HPTLC CN precoated plate using an ion-pair reversed-phase chromatographic system with tetraethylammonium chloride as the ion-pair reagent.

The newest product in the area of hydrophilic-modified precoated layers represents the HPTLC precoated plate DIOL F 254s. Because of the diol functions chemically bonded to the surface of a silica gel matrix this sorbent shows a similarity to nonmodified silica gel in its chromatographic behavior. In both cases hydroxyl groups are the active centers effecting retention. There are some differences. The hydroxyl groups take the form of an alcoholic function and for nonmodified silica gel they take the form of a silanol function. A further difference is that the diol function is bonded via a spacer group to the silica gel matrix. The chromatographic behavior of the diol sorbent is additionally influenced by this spacer group. The distinctions concerning the chemical properties of the hydroxyl groups mentioned manifest themselves in their different surface activities.

To demonstrate these differences of activity, which also influence the chromatographic properties, the chromatograms of two dyestuffs on an HPTLC precoated plate silica gel 60 and on an HPTLC precoated plate diol are compared in Figure 3.31. In the upper row there are two chromatograms

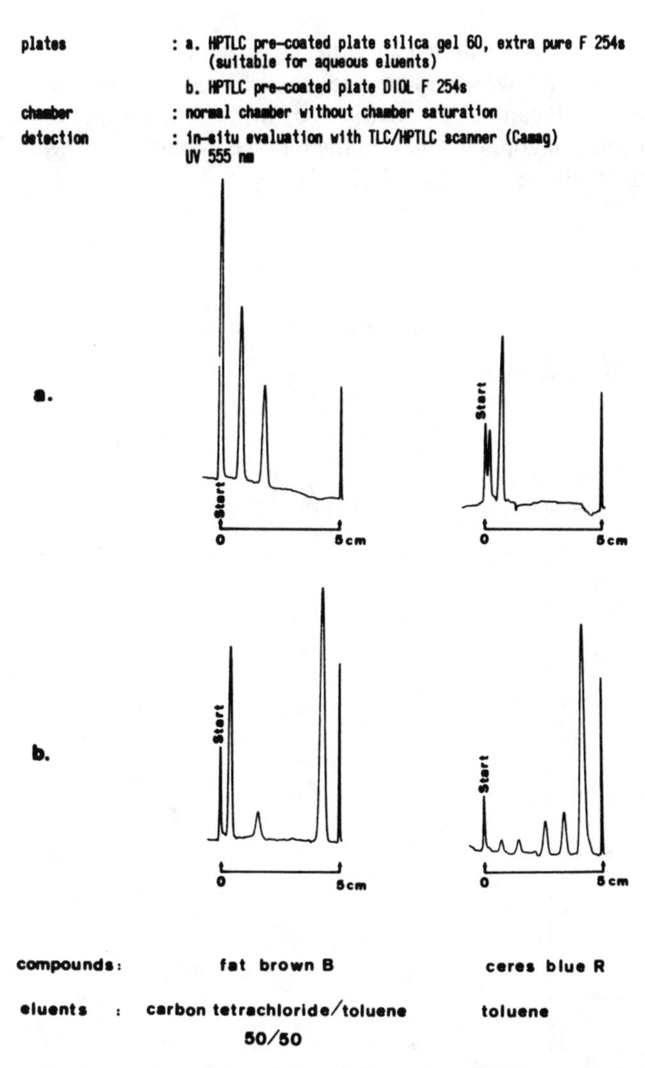

Figure 3.31. Comparison of the polarity of silica gel and DIOL precoated plates.

on the nonmodified silica gel 60 plate and below are two chromatograms on diol layers. Both chromatograms on the left side were obtained using carbon–tetrachloride and toluene (50:50) as mobile phase. The mobile phase of the two chromatograms on the right side consisted of pure toluene. Independent of the eluent used here the activity and, therefore, the retention is obviously stronger on the nonmodified silica gel than on the diol-modified sorbent. (6)

The weaker surface activity of the layer with diol modification is also expressed by its lower tendency to adsorb water from the surrounding atmosphere. This can be demonstrated by means of the dependence of the R_f values of some oligophenylenes on the relative humidity in a typical straight-phase system shown in Figure 3.32. (7)

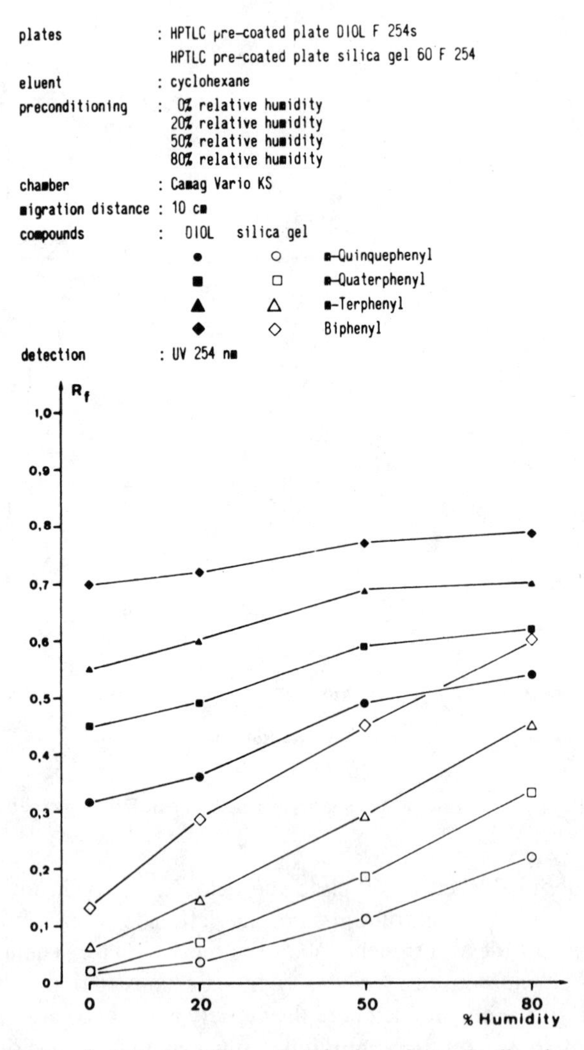

Figure 3.32. Influence of relative humidity on the Rf-values of some D-oligophenyienes on "diol" modified layers.

With the identical solvent cyclohexane and identical values of relative humidity, generally the R_f values on the diol phase are higher than those of the nonmodified silica gel. Moreover, the dependence of the strength of retention on the relative humidity is less pronounced on the diol layer, which results in smaller slopes of the curves of the R_f values in the case of the diol compared with those of the silica gel 60 layer. On the diol modified and the nonmodified silica gel 60 plates the same retention mechanisms with different intensities seem apparent. Therefore, the same or comparable mobile phases can be used for both types of precoated layers.

With regard to the properties of polarity it can be stated that the diol and the cyano phases have similar characteristics. That means they are dependent on the eluent system and on the test compounds used to make comparisons. The following figures demonstrate some applications of HPTLC precoated DIOL F 254s layers.

plate	: HPTLC pre-coated plate DIOL F 254s
eluent	: di-iso-propyl ether/acetic acid glacial 100/1 (v/v)
migration distance	: 7 cm
chamber	: normal chamber without chamber saturation
compounds	: 1 19-Nortestosterone 2 Medroxyprogesterone 3 Progesterone 4 17β-Estradiol 5 17α-Ethinylestradiol 6 meso-Hexestrol (all 0,1%)
application volume	: 300 nl
detection	: spray reagent: $MnCl_2$ - sulfuric acid heating up to 110° C for 5 min. in-situ evaluation with TLC/HPTLC scanner (Camag) UV 366 nm

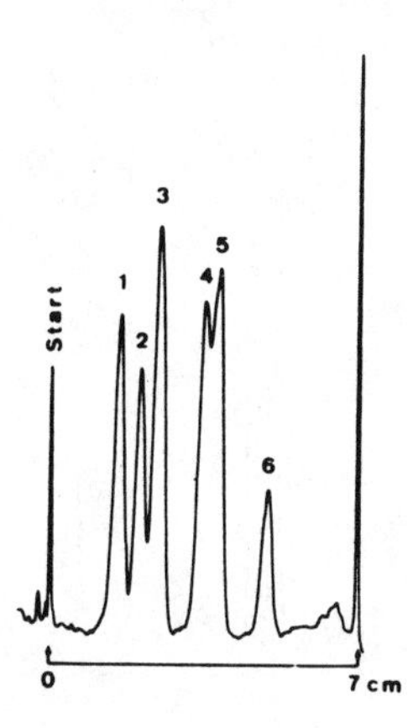

Figure 3.33. Separation of anabolic compounds.

An acidic mobile phase with the ingredients di-isopropyl ether and glacial acetic acid was used for the separation of some anabolic compounds in Figure 3.33. (8) The substances separated here are 19-nortestosterone, medroxyprogesterone, progesterone, 17β-estradiol, 17α-ethinylestradiol, and *meso*-hexestrol. Figure 3.34 demonstrates the separation of ingredients in contraceptive pills. In this case the mobile phase consists of petroleum benzene and acetone.

In conclusion with the expansion of selectivity with the introduction of the diol-modified plates the relative low dependence of the retention characteristics of these plates on the relative humidity compared with the large influence of the water content in the gas phase in case of the nonmodified silica gel layers has to be pointed out. Therefore, it is a little bit easier to get reproducible retention data on the diol plate than on the silica gel plates. All the TLC and HPTLC precoated layers that have been under discussion here are far from being the final stage in the development of thin-layer chromatography. On the contrary, it is precisely these developments that have opened

plate	: HPTLC pre-coated plate DIOL F 254s
eluent	: petroleum benzine (40–60°)/acetone 80/20 (v/v)
migration distance	: 7 cm
chamber	: normal chamber without chamber saturation
compounds	: 1 Norethindrone 2 Mestranol (all 0,2%)
application volume	: 200 nl
detection	: in-situ evaluation with TLC/HPTLC scanner (Camag) UV 254 nm

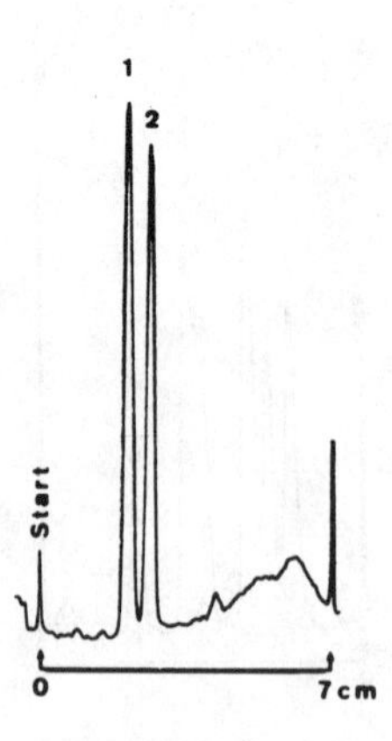

Figure 3.34. Separation of ingredients in contraceptic pills.

up new aspects for further products and application fields for qualitative and quantitative thin-layer chromatography.

REFERENCES

1. W. Jost, H. E. Hauck, and F. Eisenbeiss, *Kontakte (Darmstadt)*, p. 45 (1984).
2. H. E. Hauck and W. Jost, *Am. Lab*, 15: 72 (1983).
3. W. Jost and H. E. Hauck, *Anal. Biochem*, 135: 125 (1983).
4. Ibid., p. 126.
5. W. Jost, H. E. Hauck, and W. Fischer, *Chromatographia 21*, 377 (1986).
6. W. Jost and H. E. Hauck, *Instrumental High Performance Thin Layer Chromatography*, S. H. Traitler, A. Studer, and R. E. Kaiser (eds.). Institute for Chromatography, p. 245 (1987).
7. W. Jost and H. E. Hauck, *Chromatographia, Supp 3*, 72 (1988).
8. W. Jost and H. E. Hauck, *Instrumental High Performance Thin Layer Chromatography*, S. H. Traitler, A. Studer, and R. E. Kaiser (eds.). Institute for Chromatography, p. 248 (1987).

CHAPTER

4

ANALYSIS OF BUTYRIC ACID IN A MODEL SYSTEM

ELAINE HEILWEIL, EDWARD T. BUTTS, FIONA M. CLARK, AND WARREN E. SCHWARTZ

With the growth of the biotechnology industry there is increasing need for the economic recovery and concentration of target compounds derived from fermentation processes (1). Ion-exchange chromatography is one possible approach. A model system was designed to demonstrate the utility of ion-exchange cellulose for scavenging charged organic or biochemical species from such fermentation feedstocks. Dilute aqueous butyric acid was employed as a model feedstock, at a pH and concentration typical of production systems. Elution of the target compound was monitored by thin-layer chromatography and recovery calculated from high-performance liquid chromatography and TLC–densitometric analyses of the eluted fractions.

EXPERIMENTAL

Reagents

Potassium phosphate monobasic, glacial acetic acid, sodium chloride, sodium hydroxide, and concentrated ammonium hydroxide were reagent grade (J.T. Baker Chemical Company, Phillipsburg, NJ). Methanol was glass-distilled Omni-Solv (E. M. Science, Cherry Hill, NJ). Ethylacetate was ACS certified (Fisher Scientific, Fair Lawn NJ). *n*-Butyric acid (99%) and methyl red were purchased from Sigma Chemical Company, St. Louis, MO. The above materials were used as received. DE52 (diethylaminoethyl)-cellulose (Whatman Biosystems, Clifton, NJ) was converted to the chloride form before use.

Methods and Equipment

DE52 (diethylaminoethyl)cellulose, an anion exchange cellulose medium with a small ion capacity of 1.0 meq/dry gram, was used to scavenge butyric acid from a dilute aqueous solution. A batch absorption technique, employing equimolar quantities of a 1-wt% solution of butyric acid at pH 7.0 and DE52 converted to the chloride form at pH 7.0 was used. 100 mL of 1% butyric acid and 57 g of DE52 (79% water content) were combined and stirred at 300 rpm for 15 min using a Dataplate magnetic stirrer (Whatman, Inc. Clifton, NJ). The medium was vacuum filtered using Whatman grade 54 paper. The filtrate was retained for butyrate determination. A portion of the butyrate-laden DE52 was dispersed in deionized water and settle-packed in a conventional glass column with dual adjustable flow adaptors. A bed of 4.2 cm length × 1.6 cm diameter (8.4 mL volume) was formed. The eluate was collected during column packing. Desorption was carried out using 0.3 M acetic acid as eluant. A pumped flow system was employed (Watson Marlowe peristaltic pump model 501V, Watson Marlow, Limited, Falmouth, England), at a flow rate of 1.2 mL/min or 36 cm/h linear velocity. One-minute fractions were collected during elution with the Redirac model 2112 fraction collector (LKB Instruments Inc., Gaithersburg, MD). Eluate pH and butyrate concentration were determined. (See Figure 4.1 for profiles.)

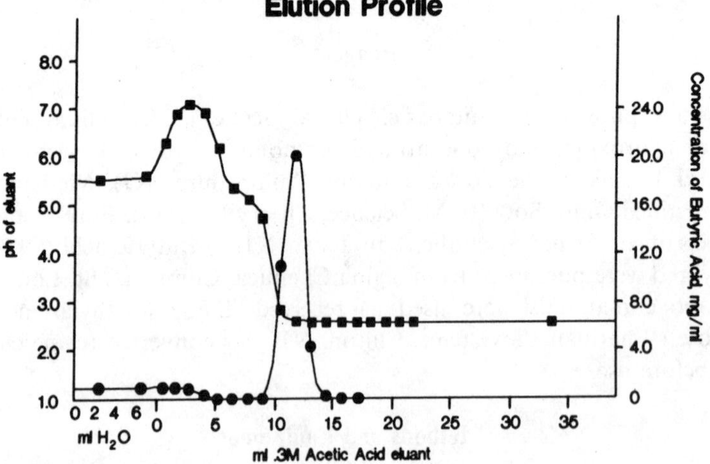

Figure 4.1. Elution of butyric acid from DEAE cellulose using 0.3 M acetic acid eluent. Column dimensions: 1.6 cm i.d. × 4.2 cm.

High-Performance Liquid Chromatography (HPLC)

A Whatman HPLC Cartridge System with a Partisphere-5 C-18 cartridge (110 mm × 4.6 mm i.d.) was used for the HPLC analysis. The mobile phase consisted of a mixture of 80% of 1.5% potassium phosphate monobasic (adjusted to pH 2.2 with 3.75 M phosphoric acid) and 20% methanol. The flow rate at ambient column temperature was 1.0 mL/min with typical column pressure in the range of 1000–1200 psi. The chromatographic system consisted of an LDC/Milton Roy Constametric IIG pump, a Valco injector valve and a Waters Differential Refractometer model R401. Peak data were obtained with an IBM PC using Nelson Analytical software. Results were calculated using the external standard method from a five-level standard curve. Duplicate 10-μL injections were made for all standards and samples.

Thin-Layer Chromatography (TLC)

TLC (2,3) was performed on the Diamond K6F 20 × 20-cm silica gel plate (Whatman, Inc., Clifton, NJ). Sample and standard solutions were applied to the layer with Drummond microcaps or a Soccorex disposable-tip, variable-volume micropipettor at $\frac{1}{2}$-in. intervals. Spot size was kept to a maximum of 0.5 cm diameter, and the spots were dried in a current of warm air. Plates were developed to a distance of 10 cm in a mobile phase consisting of ethylacetate:methanol:water:concentrated ammonia (65:25:10:5) in a paper-lined chamber equilibrated for 15 min prior to development. Following development, plates were dried at 75 °C for 15 min. After cooling to room temperature, bands were visualized by dipping the plates in a solution of methyl red, 0.04% in methanol, using a Desaga dipping chamber (Whatman, Inc., Clifton, NJ) and heating in a 105 °C oven for 15 min to develop full color. Ammonium salts of the acids appear as red bands on a yellow-orange background. After color development, chromatograms were covered with clear glass plates to protect from atmospheric carbon dioxide and prevent background darkening. Densitometry was performed immediately after the visualization step. Glass-covered chromatograms were sealed in plastic bags and stored in a freezer for photographic documentation.

TLC Densitometry

Densitometric scans of the chromatograms were obtained using the Kratos model SD3000 Spectrodensitometer. The chromatograms were scanned at 480 nm in the double-beam transmission mode, at 90° to the direction of development, that is, horizontally across the plate (4, 5). Duplicate scans were done for each chromatogram. Integration was performed with the

Spectra-Physics SP-4100 programmable integrator. A series of standards were chromatographed along with samples on each plate. Standard curves were plotted using the IBM PC and Nelson Analytical software.

RESULTS AND DISCUSSION

TLC was found to be an invaluable tool for the preliminary evaluation of the elution pattern of butyric acid from DE52 cellulose. A visual elution profile was obtained by chromatographing those fractions of the eluate in which the odor of butyric acid was discernible as well as several fractions preceding and following them (Figure 4.2). In this manner it became possible to pinpoint the fractions of interest and minimize the number of samples that would have

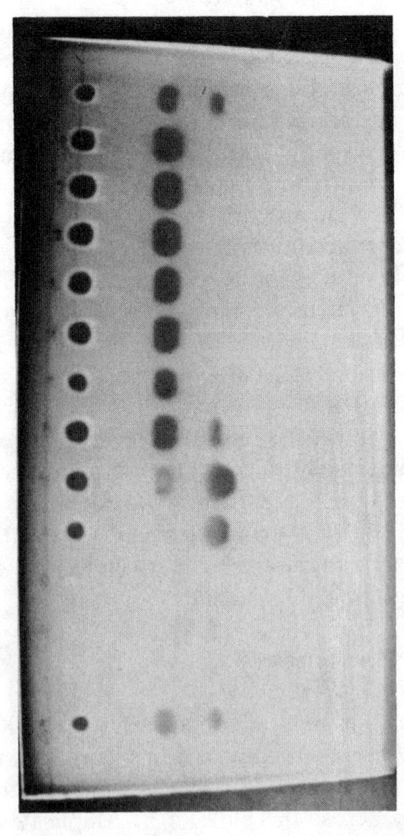

A. Fraction 17
B. Fraction 16
C. Fraction 15
D. Fraction 14
E. Fraction 13
F. Fraction 12
G. Fraction 11
H. Fraction 10
I. Fraction 5
J. Fraction 1
K. n–Butyric acid (RF 0.38) in 0.3M acetic acid (RF 0.25)

Figure 4.2. TLC Analysis of chromatographic fractions from DEAE cellulose elution.

to be quantitatively analyzed by a subsequent procedure. Quantitative analysis thus needed to be performed only on 12 out of a total of 93 fractions collected. Where the compound of interest cannot be fortuitously detected by smell as was the case here, a greater number of fractions will have to be screened by TLC, but this usually does not present great demands of time and difficulty. The complete TLC procedure is accomplished in approximately 1–2 h.

As is seen in Figure 4.2 the bulk of butyric acid elutes in fractions 11, 12, and 13, with a trace found in fraction 14. A significant amount of butyric acid is visible in fractions 1 and 5 but not in fraction 10. Subsequent TLC also showed the presence of butyric acid in fractions 1–6 with none present in fractions 7–10. The low-level presence of butyric acid in fractions 1–6 is attributable to unbound material which was entrapped in the interstitial space of the DE52 column during the uptake step but was not actually bound

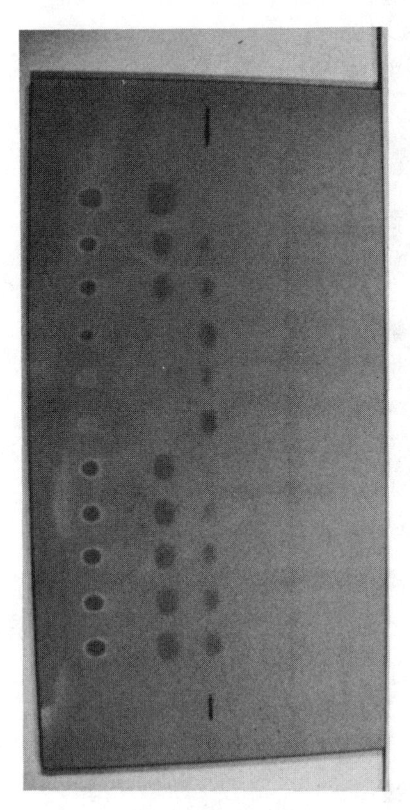

Figure 4.3. TLC chromatogram for densitometric analysis.

by the ion exchanger. The initial TLC chromatogram also shows that the acetic acid used as eluant does not actually break through the DE52 column until fraction 12 (11.0 mL of elution volume) and reaches its full concentration only in fraction 13.

Two additional chromatograms, each run in duplicate, were prepared for densitometric quantitation of the butyric acid uptake, recovery, and levels of unbound material eluting in the early fractions (Figures 4.3 and 4.4). The excellent development linearity of the Diamond K6F plate allows for scanning across the chromatogram (Figure 4.5), thus saving a significant amount of time in instrument realignment during densitometric measurement. Scanning in the double-beam mode compensates for background noise inherent in the type of detection used. Linear detector response in the range of interest of 3–30 μg was obtained (Figure 4.6), with an average correlation coefficient of

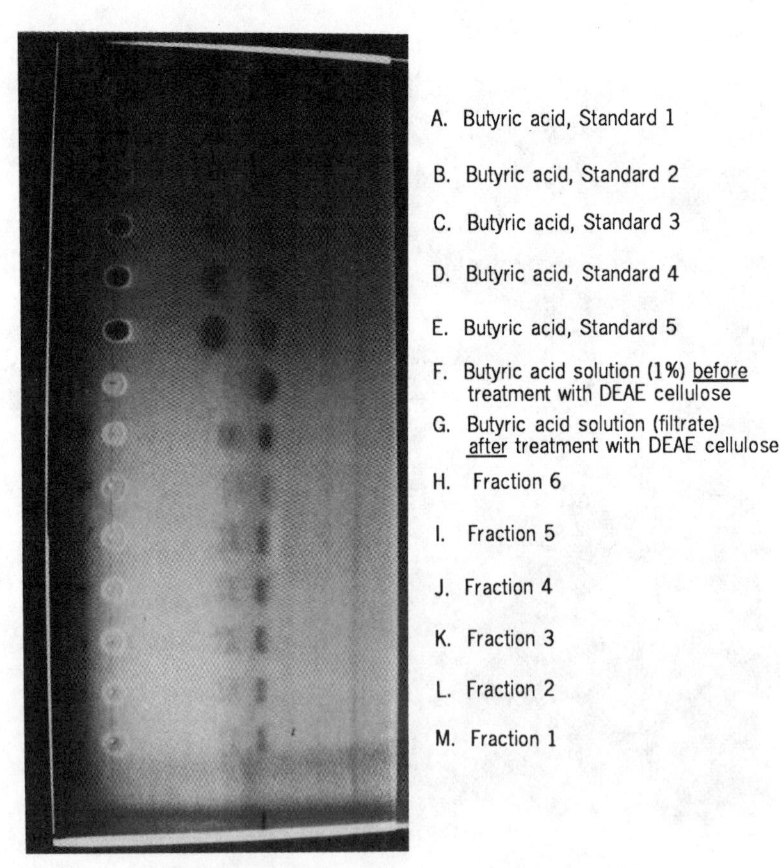

Figure 4.4. TLC chromatogram for densitometric analysis.

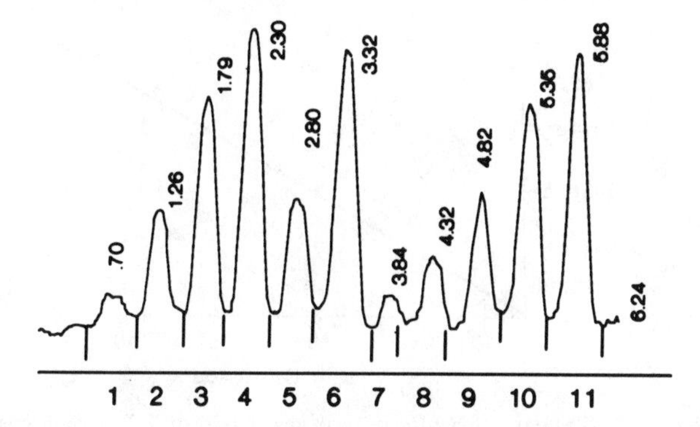

Figure 4.5. Densitometric analysis of TLC chromatogram. 1, Fraction 14; 2, fraction 13; 3, fraction 12; 4, fraction 11; 5, butyric acid solution (filtrate), after treatment with DEAE cellulose; 6, butyric acid solution (1%), before treatment with DEAE cellulose; 7, butyric acid, standard 1; 8, butyric acid, standard 2; 9, butyric acid, standard 3; 10, butyric acid, standard 4; 11, butyric acid, standard 5.

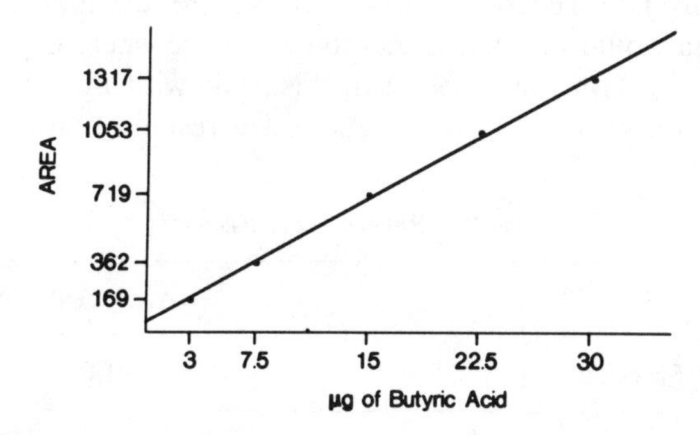

Figure 4.6. Standard curve for butyric acid based on densitometric measurement of TLC analysis.

0.9944. Sensitivity of detection for butyric acid with this method was found to be 2–3 μg, which compares favorably to the 3- to 5-μg detection limit obtained under ideal conditions with HPLC, using refractive index detection. The detection sensitivity for both TLC and HPLC could be greatly enhanced by derivatization of the acid (6) but at a great cost of in time, labor, and reagents. For the purposes of this study derivatization was not judged necessary.

High-performance liquid chromatographic quantitation (7,8) of fractions 1–6 and 10–15 was performed using a reversed–phase Partisphere-C18 (5-μm spherical, octadecyl bonded silica) cartridge system. At a flow rate of 1.0 mL/min butyric acid eluted at 3.80 min. The acetic acid used as eluant,

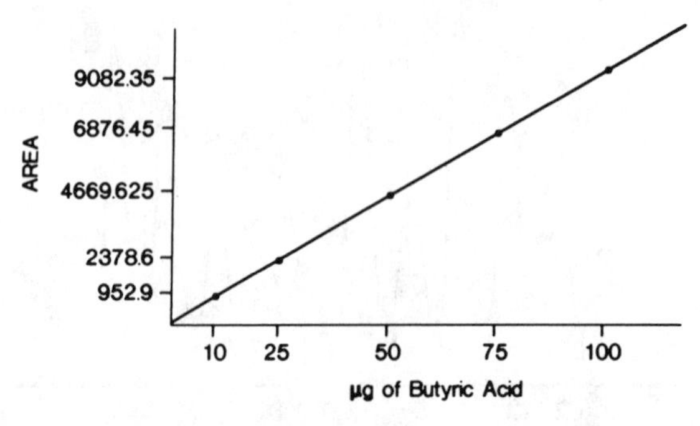

Figure 4.7. Standard curve for butyric acid based on HPLC analysis.

when present in the fractions analyzed, eluted near the void volume. The total analysis time per injection was under 5 min. Excellent linearity of detector response was obtained in the range between 10 and 100 μg of butyric acid, with a correlation coefficient of 0.9998 (Figure 4.7). Normally refractive index detection would not be the method of choice where high sensitivity is required. However, for the purposes of this study with TLC confirmation of HPLC results, the sensitivity was adequate. The results obtained from HPLC

Table 4.1. Butyric Acid Recovery

	Analytical Method	
Sample/Fraction	HPLC	TLC
1	2.70	2.80
2	2.55	2.27
3	0.78	0.73
4	0.75	0.87
5	1.02	1.26
6	0.54	0.98
11	14.85	13.93
12	27.19	24.75
13	5.77	5.10
14	0.37	0.37
mg recovered (fractions 11–14)	48.18	44.15
mg unbound (fractions 1–6)	8.34	8.91
Total uptake (mg)	63.80	61.35
% Recovery (based on total uptake)	75.5	72.0
% Recovery of bound butyric acid	86.9	84.2

analysis are in good agreement with TLC–densitometry (Table 4.1), so that either method can be applied, depending on availability in the laboratory. TLC, however, is the method of choice for rapid screening and gaining a quick visual profile of an elution pattern.

CONCLUSION

A method for the isolation of an organic acid using ion-exchange cellulose along with procedures for monitoring the elution pattern and analysis of the eluate by TLC and HPLC was presented.

The target compound in this system was recovered from dilute aqueous solution and concentrated using anion exchange chromatography. High recoveries from the anion exchange column have been demonstrated, the target being recovered in a small volume in a more concentrated form.

This model system demonstrates the utility of ion-exchange *cellulose media* in the recovery and concentration of charged biological molecules from dilute aqueous solution, as in fermentation feedstocks.

In our laboratory the HPLC procedure was the primary quantitative analysis method, but results show that quantitation by TLC yields comparable results and can be used interchangeably. Thin-layer chromatography is ideally suited to monitoring the elution pattern and is invaluable when used as complementary methodology along with HPLC or any other subsequent analytical procedure.

REFERENCES

1. D. J. Fink, L. M. Curran, and B. R. Allen *Research Needs in Non-Conventional Bioprocess*, Batelle, Columbus, OH, 1985.
2. E. Stahl, *Thin Layer Chromatography*, 2d ed. Springer-Verlag, Berlin, 1969.
3. J. G– Kirchner, *Thin Layer Chromatography*, 2d ed. Wiley-Interscience, New York, 1978.
4. J. Touchstone and J. Sherma, *Densitometry in Thin Layer Chromatography*, Wiley, New York, 1979.
5. L.R. Treiber, *Quantitative Thin Layer Chromatography and Its Industrial Applications*, Chromatographic Science Series, Vol. 36. Dekker, New York, 1987..
6. E. Heilweil and T. E. Beesley, *J. Liq. Chromatogr.* 4: 2193 (1981).
7. P. R. Brown, *High Pressure Liquid Chromatography—Bio-Chemical and Bio-medical Applications*, Academic, New York, 1973.
8. P. A. Bristow. *Liquid Chromatography in Practice*, hetp PRESS, Wilmslow, Cheshire, UK, 1976.

CHAPTER

5

TLC-IMMUNOSTAINING OF GLYCOLIPIDS

MEGUMI SAITO AND ROBERT K. YU

Thin-layer chromatographic (TLC)-immunostaining is a method for detecting glycolipids separated on TLC plates using specific anti-glycolipid antibodies. It was originally developed by Magnani et al. (1) who used ^{125}I-labelled cholera toxin as the ligand to detect the ganglioside GM1. After developing the ganglioside on plastic-backed silica gel plates, they treated the plates with a polyvinylpyrrolidone solution to make a polymer film on the surface of the silica gel layer so that the binding reaction of the toxin and GM1 could proceed without the silica gel coming off the plates. ^{125}I-Labeled cholera toxin bound to GM1 was then detected by autoradiography. Thereafter, this glycolipid-overlay technique has proved to be applicable to detection of essentially all kinds of glycolipids by using specific antibodies instead of the toxin, and has become widely used in glycolipid research.

METHOD

TLC-immunostaining is composed of two essential steps: separation of glycolipids on TLC, and in situ detection of the glycolipids with specific antibodies. Since TLC and high-performance TLC (HPTLC) have a high resolution for the separation of glycolipids, minor components that are difficult to purify can be separated and characterized by HPTLC, since the use of specific anti-glycolipid antibodies makes it possible to detect glycolipids with strict structural specificity. The sensitivity of this method is generally very high, nanogram-order or less amounts of glycolipids being sufficient. The original technique reported by Magnani et al. has been modified in many respects to improve the stability and reproducibility of the method. Glass-backed silica gel plates, which generally bring finer separation of glycolipids but lose silica gel more easily during incubation than plastic-backed ones, were successfully used by treating them with another organic polymer, polyisobutylmethacrylate, at higher concentrations (2, 3). To avoid the use of radioactive materials and save time spent for exposure of plates to X-ray films, various types of immunoenzyme methods, such as immunoperoxidase techniques (4, 5) or avidin-biotin enzyme systems (6), have been developed without significant reduction in the sensitivity of the method. In addition to cholera toxin and anti-glycolipid antibodies, other kinds of ligands, such as lectins, have also been used (7). Such ligands, having less strict specificity compared to antibodies, would be especially useful for obtaining information about glycolipids of unknown structures.

Figure 5.1 shows two different procedures of TLC-immunostaining used in our laboratory; one includes an autoradiographic step (8) and the other an immunoperoxidase technique (9). (For more details, see the section Analysis of in situ reactions of glycolipids.)

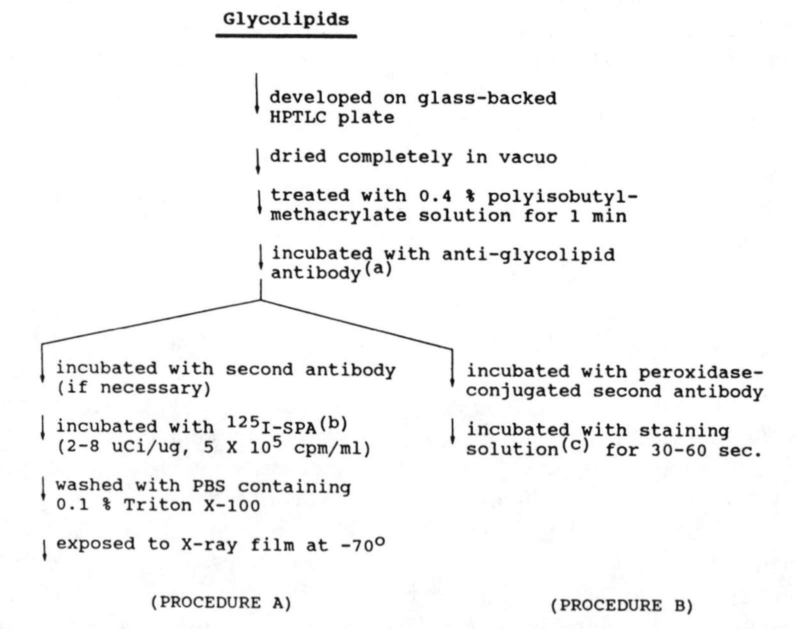

Figure 5.1. Procedures of TLC-immunostaining. (*a*) Each incubation is usually carried out at room temperature for 1 h, followed by washing with phosphate-buffered saline (PBS), unless specified. (*b*) SPA, staphylococcal protein A. (c) The staining solution is prepared by mixing solutions of 3,3'-diaminobenzidine tetrahydrochloride (2 mg/mL of PBS, 10 mL) and imidazole (2 mg/mL of PBS, 10 mL), and hydrogen peroxide (50%, 5 μL).

IDENTIFICATION AND CHARACTERIZATION OF GLYCOLIPIDS AND ANTI-GLYCOLIPID ANTIBODIES

TLC-immunostaining is thus a very powerful method for examining specificities of anti-glycolipid antibodies and screening for their presence in body fluids or culture media. It has also been vigorously employed for identification and characterization of "tumour-specific" and "differentiation-dependent" glycolipid antigens (10,11), including the sialosyl-Le^a antigen (12,13), GD3 (14), gangliosides having the sialosyl(2-6)galactosyl residue (15), globotriaosylceramide (16), poly-X antigens (17), and GD2 (8). Specificities of antibodies detected in other pathological conditions have also been examined by this method. For example, autoantibodies, which are detected in sera of patients with demyelinating neuropathy and plasma cell dyscrasia, have been shown to react with certain acidic glycolipids (18, 19). Since the antibodies are known to interact with myelin-associated glycoproteins (MAG) (20), it was thought that the glycolipids and MAG share the same antigenic determinant(s). Structural analysis has revealed that one major glycolipid is sulfated glucuronyl neolactotetraosyl ceramide (sulfated glucuronyl paragloboside, SGPG) (21, 22) and the other is sulfated glucuronyl lactosaminyl paragloboside (SGLPG) (22, 23).

Antibodies directed toward normal cells or tissues, including A2B5 (24) or 18B8 (25), have also been analyzed by this method.

QUANTITATION OF GLYCOLIPIDS

In binding reactions between ligands and glycolipids on TLC plates, the amount of a ligand bound to a glycolipid is assumed to correlate with that of the glycolipid applied to the plate. This idea led to the development of quantitative analytical methods for glycolipids using the TLC-immunostaining technique. The quantitation of lipids is usually achieved by scanning autoradiograms of plates or chromatograms stained with immunoenzyme techniques. Kohriyama et al. (26) devised a quantitative immunostaining method by scanning autoradiograms, and analyzed the subcellular distribution of sulfated glucuronic acid-containing glycolipids, SGPG and SGLPG, using this method.

The subcellular fractionation of bovine spinal accessory nerves was carried out by continuous sucrose density centrifugations (27). Samples of acidic lipids for HPTLC were prepared as follows. The total lipids were extracted from subcellular fractions with chloroform/methanol/water (30:60:8). The acidic lipids were isolated from the total lipids by DEAE–Sephadex column chromatography (A-25, acetate form) using chloroform/methanol/1.2 M sodium acetate (30:60:8) as an eluent. The sulfated glucuronyl glycolipids used as external standards for quantitation were purified from bovine cauda equina (23). TLC-immunostaining was carried out according to the method shown in Figure 5.1, procedure A. The acidic lipid samples and different amounts of the standard glycolipids were developed on HPTLC. The sulfated glucuronyl glycolipids, SGPG and SGLPG, separated on the plate were detected with the serum of a patient with neuropathy and IgM M-proteins. The plate was treated with anti-human goat immunoglobulin IgG and ^{125}I-labeled SPA, successively, and the autoradiogram obtained was scanned by a densitometer equipped with a data processor. The amount of the glycolipids in each sample was computed based on the standard curves, which were obtained with the standard glycolipids developed on the same plate with the

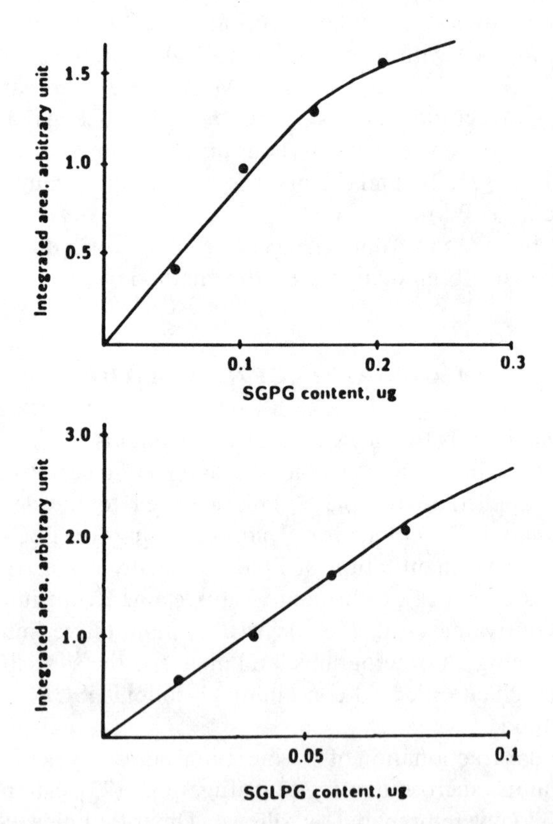

Figure 5.2. Standard curves for sulfated glucuronyl glycolipids, SGPG and SGLPG. Different amounts of each lipid were stained and scanned using the method described in the text. Under the condition used, the patient's serum used in obtaining the standard curves had about three times higher affinity for SGLPG than SGPG.

samples. Figure 5.2 shows the standard curves for SGPG and SGLPG. Under the conditions used, the dose–response curves were essentially linear up to 50 ng for SGPG and up to 15 ng for SGLPG. The subcellular distribution of SGPG and SGLPG in bovine spinal accessory nerves, relative to protein, acetylcholine esterase and 2′,3′-cyclic nucleotide 3′-phosphodiesterase, is shown in Table 5.1. The distribution patterns of these glycolipids were similar and the highest specific concentration of each glycolipid was found in the axolemma-enriched fraction, suggesting their close association with the same constituents.

Densitometry of chromatograms stained using immunoenzyme techniques has also been successfully applied to the quantitation of glycolipids, including Hangenutzui–Deicher antigens (28).

ANALYSIS OF IN SITU REACTION PRODUCTS OF GLYCOLIPIDS

TLC-immunostaining indicates that glycolipids developed on TLC plates can react with ligands such as antibodies and suggests that they can also be subjected to reactions other than binding with the ligands. This possibility has been confirmed by several authors by carrying out various types of reactions involving glycolipids on TLC plates. We examined the effect of neuraminidase on gangliosides developed on silica gel plates, and found that in the absence of detergent the hydrolytic reaction of gangliosides proceeded more easily on plates than in solution (3). Based on this finding, we developed a new analytical method for determining the carbohydrate structures of gangliosides. In principle, one can modify unknown glycolipid structures by enzymatic or chemical means after separation on TLC plates and identify the structures of the glycolipid remnants by specific antibodies. Once this is done, it is possible to deduce the structures of the parent glycolipids based on the specificity of enzymatic or chemical reactions. In this method, gangliosides are modified by *Arthrobacter ureafaciens* neuraminidase on the plate, and the generated asialogangliosides are treated with an anti-asialo GM1 (anti-Gg4) antibody. If a ganglioside of unknown structure can be detected with the antibody after treatment with the enzyme, it means that the ganglioside has the same carbohydrate core structure as Gg4. Details of the method are as follows: Human brain gangliosides (1.25 g as sialic acid) were applied to a HPTLC glass-backed plate (Silica Gel 60, E. Merck) in 5-mm streaks, and developed to 7 cm from the bottom edge of the plate with chloroform–methanol–0.22% aq. $CaCl_2 \cdot 2H_2O$ (50:45:10). After drying in vacuo for at least 2 h, the plate was dipped in a 0.4% polyisobutylmethacrylate solution for 1 min. The polymer solution was prepared by diluting a 2.5% chloroform solution with *n*-hexane. Each lane of the plate was overlayed with an *A. ureafaciens* neuraminidase solution (10, 20, and 40 mU/mL of 0.1 M sodium acetate, pH 4.8, about 0.5 mL per lane), and incubated at room temperature for 2 h in a humid plastic box. The plate was washed with phosphate-buffered saline (PBS) and dried briefly in air. Thereafter, the plate was treated with an anti-Gg4 rabbit serum, which was diluted with 0.3% gelatin–PBS, for 1h, and with ^{125}I-labeled staphylococcal protein A (about 5 $\times 10^5$ cpm/mL of 0.3% gelatin–PBS) for another hour, successively. Finally, the plate was thoroughly washed with 0.1% Triton X-100 in PBS, and exposed to an X-ray film overnight. As shown in Figure 5.3 the anti-Gg4 antibody reacted exclusively with Gg4 (lane 5) but not with any intact gangliosides (lane 4). After treatment with *A. ureafaciens* neuraminidase, gangliosides such as GM1, GD1a, GD1b, GT1a, GT1b, and GQ1b became reactive with the antibody, indicating that these gangliosides had the same carbohydrate core structure as Gg4, which was revealed by hydrolysis (lanes 1 to 3). The intensity of each positive band increased in proportion to the concentration of the enzyme used.

This method was applied to the analysis of gangliosides of PC 12 cultured pheochromocytoma cells (29). The ganglioside composition of PC 12 cells

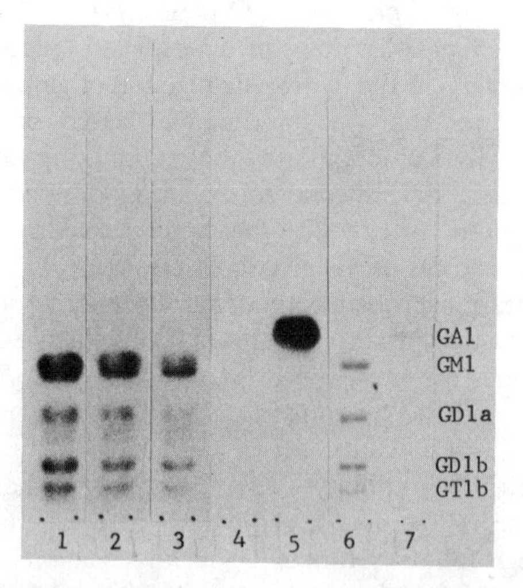

Figure 5.3. Detection of gangliosides having the same carbohydrate core structure as Gg4 in human brain gangliosides. The ganglioside mixture (lanes 1 to 4 and 6) and Gg4 (lanes 5 and 7) were developed on a HPTLC plate. Lanes 1 to 5 were incubated with 40 mU/mL (lane 1), 20 mU/mL (lane 2), or 10 mU/mL (lane 3) of an *A. ureafaciens* neuraminidase solution, or with the buffer (0.1 M sodium acetate, pH 4.8, lanes 4 and 5), followed by incubation with anti-Gg4 antibody and ^{125}I-labeled SPA, successively. Lanes 6 and 7 were visualized by the α-naphthol–sulfuric acid reagent.

had not been well characterized although some of them were known to be fucose-containing glycolipids (30). Total gangliosides extracted from cells were separated into monosialo-, disialo-, and tri- and tetrasialo fractions by DEAE–Sephadex column chromatography, and were subjected to analysis by this method. As shown in Figure 5.4 six major ganglioside bands were visualized with the α-naphthol–sulfuric acid reagent: three in the monosialo fraction, one in the disialo fraction, and two in the tri- and tetrasialo fraction. After in situ treatment of gangliosides with the enzyme, many gangliosides, including two major ones in the tri- and tetrasialo fraction, became reactive with the antibody while major gangliosides in the mono- and disialo fractions were not detected by the antibody. This result clearly demonstrated that PC 12 cells had many kinds of "brain-type" gangliotetraosyl-series gangliosides.

Various combinations of enzymes and ligands can be used in this method, providing different information about the parent glycolipids. When GD1b was treated with jack bean β-galactosidase (1.875 U/mL of 0.1 M citrate–0.2% taurodeoxycholate, pH 4.5) on TLC plates, it was converted to

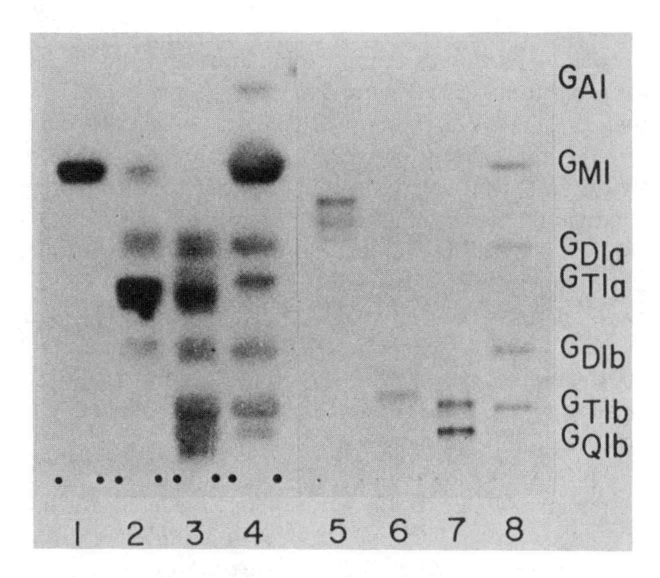

Figure 5.4. In situ treatment of gangliosides isolated from PC 12 cells with neuraminidase. A total ganglioside was separated into monosialo-, disialo-, and tri- and tetrasialo fractions by DEAE–Sephadex column chromatography. Each fraction was developed on a HPTLC plate (lanes 1 and 5 were the monosialo fraction, lanes 2 and 6 the disialo fraction, and lanes 3 and 7 the tri- tetrasialo fraction). A human brain ganglioside mixture was developed on lanes 4 and 8. Lanes 1 to 4 were stained by anti-Gg4 antibody after treatment with *A. ureafaciens* neuraminidase (40 mU/mL of 0.1 M sodium acetate, pH 4.8). Lanes 5 to 8 were visualized by the α-naphthol–sulfuric acid reagent.

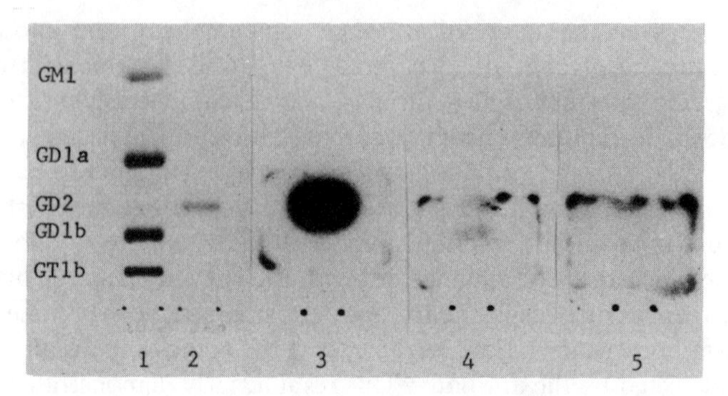

Figure 5.5. Conversion of the ganglioside GDlb to GD2 by in situ treatment with jack bean β-galactosidase. A human brain ganglioside mixture (lane 1), purified GD2 (lanes 2 and 3), and purified GDlb (lane 4 and 5) were developed on a HPTLC plate. Lanes 1 and 2 were visualized by the α-napthol–sulfuric acid reagent. Lane 3 was directly stained with anti-GD2 antibody, 3F8. Lane 4 was incubated with β-galactosidase (1.875 U/mL of 0.1 M citrate–0.2% taurodeoxycholate, pH 4.5) at room temperature for 2 h, followed by incubation with anti-GD2 antibody and ^{125}I-labeled SPA. Lane 5 was incubated with the buffer instead of the enzyme solution.

GD2, which was detected with a monoclonal antibody, 3F8, directed toward cultured neuroblastoma cells (Figure 5.5).

Biosynthetic reactions of glycolipids can also be carried out on plates. Samuelsson (31) reported that blood group glycolipids could be fucosylated on TLC plates. Total neutral glycosphingolipids, which were prepared from plasma of individual blood donors, were separated on HPTLC. Each lane developed was overlayed with an incubation medium, which was composed of 650 μL of an enzyme preparation, about 1.3×10^6 cpm of GDP-L-[^{14}C]fucose, 25 μmol Tris–HCl (pH 7.5), 10 μmol NaN_3, 10 μmol ATP, 7.5 μmol $MgCl_2$, and 0.1% Triton X-100. For the enzyme preparation, a microsomal fraction obtained from pig intestine mucosa was used. The autoradiography of the plate showed that some glycolipids, including Le^b hexaglycosylceramide, were fucosylated.

The glycolipid-overlay technique has also been used to analyze the receptor function of glycolipids for viruses and bacteria. Bock et al. (32) examined the interaction of *E. coli* 36692 with 34 different glycolipids and reported that only glycolipids having the structure of Gal(α1–4) Gal at the internal or terminal position interacted with the bacteria on plates.

CONCLUSIONS

TLC-immunostaining is now recognized as a very useful method for identification and characterization of glycolipids as well as anti-glycolipid antibodies. In addition, it has been shown that the silica gel surface provides a unique site where various types of reactions, such as degradative or biosynthetic reactions of glycolipids, can proceed. Further application of the method in this field would provide more information about the metabolism and functions of glycolipids.

ACKNOWLEDGMENTS

A portion of the work reported here was carried out with N. Kasai and T. Kohriyama in the authors' laboratory. Financial support was provided by NIH Grants NS 11853, NS 23102, and NS 26994.

REFERENCES

1. J. L. Magnani, D. F. Smith, and V. Ginsburg, *Anal. Biochem.* 109: 399 (1980).
2. N. Kasai, M. Naiki, and R. K. Yu, *J. Biochem.* 91: 261 (1984).
3. M. Saito, N. Kasai, and R. K. Yu, *Anal. Biochem.* 148: 54 (1985).
4. H. Higashi, K. Ikuta, S. Ueda, S. Kato, Y. Hirabayashi, M. Matsumoto, and M. Naiki, *J. Biochem* 95: 785 (1984).
5. M. L. Harpin, M. J. Coulon-Molerec, P. Yeni, F. Danon, and N. Baumann, *J. Immunol. Methods* 78: 135 (1985).
6. J. Buehler and B. A. Macher, *Anal. Biochem* 158: 283 (1985).
7. T. Momoi, T. Tokunaga, and Y. Nagai, *FEBS Lett* 141: 6 (1982).
8. M. Saito, R. K. Yu, and N-K. V. Cheung, *Biochem. Biophys. Res. Commun.* 127: 1 (1985).
9. S. Kusunoki, T. Kohriyama, A. R. Pachner, N. Latov, and R. K. Yu, *Neurology* 37: 1795 (1987).
10. T. Feizi and R. A. Childs, *TIBS*, 24 (1985).
11. S. Hakomori, in *Oncogenes and Growth Factors*, R. A. Bradshaw and Z. Steplewski (Eds.). Elsevier Science, Amsterdam, 1987), pp. 218–226.
12. J. L. Magnani, M. Brockhaus, D. F. Smith, V. Ginsburg, M. Blaszczyk, K. F. Mitchell, Z. Steplewski, and H. Koprowski, *Science* 212: 55 (1981).
13. J. L. Magnani, B. Nilsson, M. Brockhous, D. Zop, Z. Steplewski, H. Koprowski, and V. Ginsburg, *J. Biol. Chem.* 257: 14 365 (1982).
14. C. S. Pukel, K. O. Lloyd, L. R. Trabassons, W. G. Dippold, H. F. Oettgen, and L. J. Old, *J. Exp. Med.* 155: 1133 (1982).
15. S. Hakomori, E. Nudelmann, S. B. Levery, and C. M. Patterson, *Biochem. Biophys. Res. Commun.* 113: 791 (1983).
16. E.Nudelman, R. Kannagi, S. Hakomori, M. Parsons, M. Lipinski, J. Wiels, M. Fellous, and T. Tursz, *Science* 220: 509 (1983).

17. Y. Fukushi, E. Nudelman, S. B. Levery, and S. Hakomori, *J. Biol. Chem.* 259: 10511 (1984).

18. A. A. Ilyas, R. H. Quarles, T. D. MacIntosh, M. L. Dobersen, B. D. Trapp, M. C. Dalakas, and R. O. Brady, *Proc. Natl. Acad. Sci. U.S.A.* 81: 1225 (1984).

19. L. Freddo, T. Ariga, M. Saito, L. J. Macala, R. K. Yu, and N. Latov, *Neurology* 35: 1420 (1985).

20. N. Latov, R. B. Gross, J. Kastleman, Y. Flanagan, S. Lemme, D. A. Alkaitis, M. R. Olarte, W. H. Sherman, L. Chess, and A. S. Penn, *N. Engl. J. Med.* 303: 618 (1980).

21. T. Ariga, M. Saito, L. Freddo, N. Latov, L. J. Macala, and R. K. Yu, *Trans. Am. Soc. Neurochem.* 16: 172 (1985).

22. P. E. Chou, A. A. Ilyas, J E. Evans, R. H. Quarles, and F. B. Jungalwala, *Biochem. Biophys. Res. Commun.* 128: 383 (1985).

23. T. Ariga, T. Kohriyama, L. Freddo, N. Latov, M. Saito, K. Kon, S. Ando, M. Suzuki, M. E. Hemling, K. L. Rinehart, S. Kusunoki, and R. K. Yu, *J. Biol Chem.* 262: 848 (1987).

24. S. K. Kundu, M. A. Pleatman, W. A. Redwine, A. E. Boyd, and D. M. Marcus, *Biochem. Biophys. Res. Commun.* 116: 836 (1983).

25. C. Dubois, J. L. Magnani, G. B. Grunwald, S. L. Spitalnik, G. D. Trisler, M. Nirenberg, and V. Ginsburg, *J. Biol. Chem.* 261: 3826 (1986).

26. T. Kohriyama, S. Kusunoki, T. Ariga, J. E. Yoshino, G. H. DeVries, N. Latov, and R. K. Yu, *J. Neurochem.* 48: 1516 (1987).

27. J. E. Yoshino, J. Griffin, and G. H. DeVries, *J. Neurochem* 41: 1126 (1983).

28. H. Higashi, Y. Fukui, S. Ueda, S. Kato, Y. Hirabayashi, Y., M. Matsumoto, and M. Naiki, *J. Biochem* 95: 1517 (1984).

29. T. Ariga, L. J. Macala, M. Saito, R. K. Margolis, L. A. Greene, R. U. Margolis, and R. K. Yu, *Biochemistry* 27: 52 (1988).

30. R. U. Margolis, M. Mazzulla, L. A. Greene, and R. K. Margolis, *FEBS Lett.* 172: 339 (1984).

31. B. E. Samuelsson, *FEBS Lett* 167: 47 (1984).

32. K. Bock, M. E. Breimer, A. Brignole, G. C. Hansson, K-A. Karlsson, G. Larson, H. Leffler, B. E. Samuelsson, N. Stromberg, C. S. Eden, and J. Thurin, *J. Biol. Chem.* 260: 8545 (1985).

CHAPTER

6

TLC IN PHARMACEUTICAL RESEARCH

NELU GRINBERG, JOHN A. BAIANO, GARY BICKER,
PATRICIA TWAY, AND DEAN ELLISON

Drug production in the pharmaceutical industry is the result of a joint effort between several fields of expertise, such as biotechnology, natural product isolation, and organic synthesis. Once a drug is produced by one of these areas it is introduced into the development scheme outlined in Table 6.1. Table 6.1 clearly emphasizes the important role of chromatography as an analytical technique. Further, it is obvious that a knowledge of the number and nature of impurities that may be present is essential in the characterization of drug purity. The increased demand for new drugs not only requires accurate assessment of purity but also dictates that it be performed in a time-efficient manner. Thin-layer chromatography (TLC) fulfills these requirements.

TLC is a powerful tool for monitoring impurity profiles of drugs not only because of its speed but also because of the wide range of detection methods available as opposed to HPLC methods. Validation of the TLC method is essential in establishing reliable methodology (1). Validation must take into account all factors that may influence the reproducibility of the final results. It is well known that the R_f values of the separated compounds are dependent upon chromatographic chamber saturation (2), stationary-phase quality and activation (3), the migration distance of the solute (4), layer thickness (5), and so forth. Other important aspects of TLC validation include proper interpretation of the chromatograms. The general tendency among chemists is to attribute multiple spots in a TLC separation to impurities (6). This may not always be the case, since some spots may be artifacts of the system due to the degradation of the sample during development. Therefore, the validation of any TLC method must also include studies to probe the system for on-plate decomposition.

In the present paper some of the parameters that must be considered to develop a TLC method are outlined. A number of applications involving drug substances are presented. Focus will be in particular on the influence of chamber saturation on solute migration on both silica-gel and RP-18 station-

Table 6.1. Analytical Physical Chemical Support Required in the Research and Development of a Pharmaceutical

Biotechnology — Natural products — Organic synthesis

Stage	Analytical physical chemical support
Pharmacological screening	Characterization 1. Mass spectrometric 2. Nuclear magnetic resonance 3. GC/MS, LC/MS 4. Chromatographic 5. Spectroscopic 6. Crystallographic 7. FTIR 8. Biotechnology A. Amino acid analyzer B. Sequence analysis
Decision to develop	
Process R and D	Physical characterization of bulk drug purity and stability 1. Spectroscopic 2. Chromatographic 3. Physical properties 4. Characterization of process impurities 5. Characterization of degradates and degradation kinetics
Safety assesment	1. Support of preparation and evaluation of simple liquid dosage forms 2. Biological fluid assays of drug levels in animal species
Pharmaceutical R and D	Characterization of drug in dosage form 1. Compatibility studies of drug and excipients 2. Development and application of drug assays to assess drug potency, uniformity, and dissolution 3. Stability kinetics of drug formulation
Clinical trials, phases I–III	Pharmacokinetics evaluation of the adsorption, distribution, metabolism, and elimination of drugs in human subjects through development and application of biological fluid assays
FDA approval	Preparation of all analytical documents to assure safety and efficacy of drug product
Quality control	Analytical testing for continued assurance of the potency, uniformity, and stability of marketed products

Table 6.2. Structures of the Model Compounds

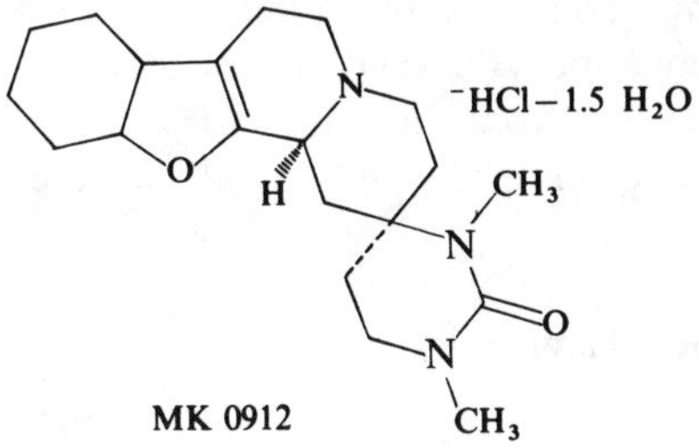

$^{-}HCl-1.5\ H_2O$

MK 0912

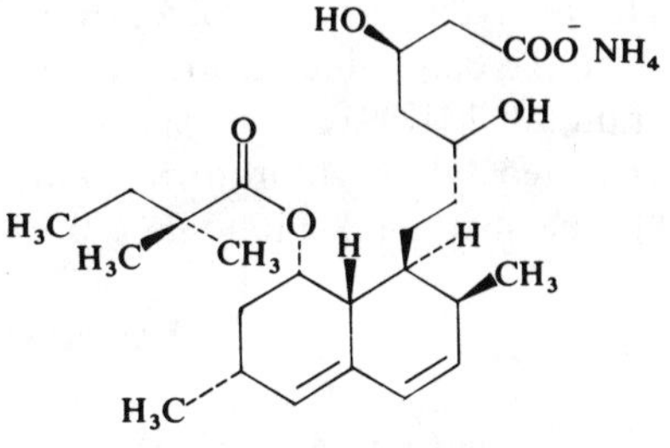

L-654, 969

$(C_2H_5)_2NCH_2CH_2O$—⟨C₆H₄⟩—$C(OH)(Ar1)CH_2$—Ar2

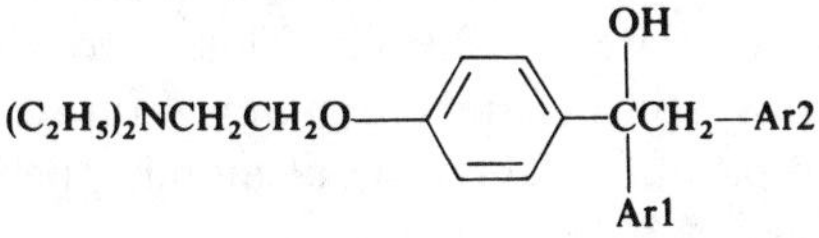

Ar = substituted aromatic

L-584, 428

ary phases, the effect of silica-gel stationary-phase activation on solute migration, and the meaning of the impurities detected in a TLC separation. L-584,428, L-654,969, and MK0912 were used as model compounds to study the above phenomena. Their structures are presented in Table 6.2.

EXPERIMENTAL

Stationary Phases

HPTLC silica-gel and RP-18 plates 10×10 cm and RP-18 5×20 and 10×20 cm were purchased from E. Merck (Darmstadt, Germany).

Mobile Phases

The mobile phases used in these experiments consisted of ethyl acetate–hexane–methanol–ammonia (50:35:15:2, v/v), methylene chloride–tetrahydrofuran–triethylamine (85:15:7, v/v), and acetonitrile–methanol– 0.1 M aqueous ammonium acetate (70:10:20, v/v). The apparent pH of the latter mobile phase was adjusted to 5.5 with glacial acetic acid.

Sample Preparation

The analytes were dissolved in methanol or methanol containing 0.05% BHT. To determine the origin of extraneous spots detected in the chromatogram of MK0912, the spots were scraped from the stationary phase, extracted with chloroform and again developed with ethyl acetate–hexane–methanol–ammonia (50:35:15:2, v/v).

Chromatographic Chamber

The chromatography was performed in rectangular chromatographic chambers. To determine the influence of chamber saturation on the separation, the tank was lined with filter paper, previously impregnated with the mobile phase, and allowed to equilibrate for 30 min. This is to ensure that the atmosphere in the chromatographic chamber remains saturated with mobile-phase vapors during the separation. The level of mobile phase was 0.5 cm.

Detection

After the chromatograms were removed from the chromatographic chambers, they were dried under a stream of nitrogen and inspected under long-wave UV (366 nm) and short-wave UV (254 nm) and the results were recorded. Following this, in the case of MK0912, the plates were placed in an oven at 120 °C for 10 min and inspected again under short- and long-wave UV and the results were again recorded. In some cases a 20% H_2SO_4 in methanol spray was used followed by 5 min heating at 120 °C. This treatment causes some analytes to fluoresce when the plate is visualized under long-wave UV(7).

RESULTS AND DISCUSSION

The Influence of Chromatographic Chamber Saturation on Resolution

Chromatographic chamber saturation plays an important role in TLC. Saturation depends on the individual vapor pressure of each of the components of the eluent. In the case of silica gel, the mobile-phase vapors are adsorbed into the stationary phase, modifying the entire surface. This phenomenon leads to different adsorption mechanisms since the nature of the solute–stationary phase interaction has been changed (2).

Several different experiments were performed to evaluate the influence of chamber saturation on the separation. The results are presented in Figures 6.

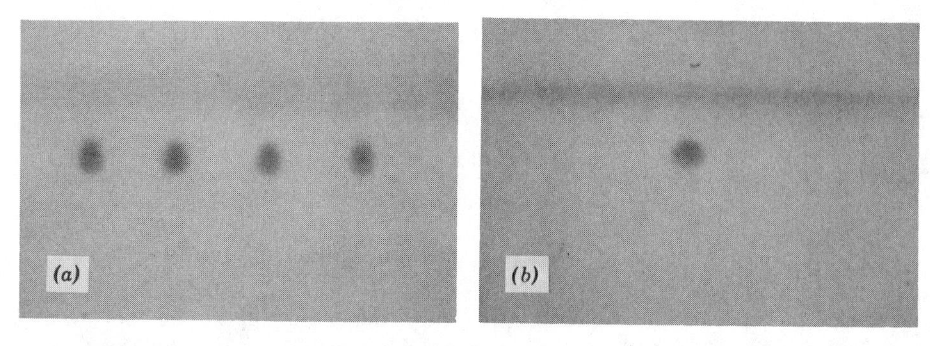

Figure 6.1. Influence of saturation of the chromatographic chamber on the separation of L-584,428 using silica-gel HPTLC. Mobile phase: methylene chloride–tetrahydrofuran–triethylamine (85:15:7, v/v). (*a*)Saturated conditions; (*b*) nonsaturated conditions.

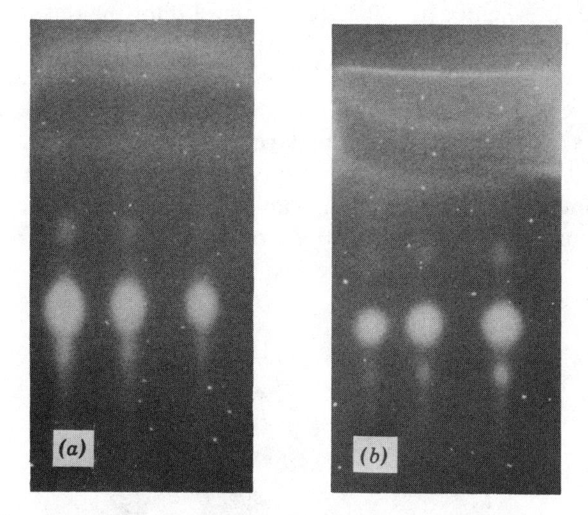

Figure 6.2. Influence of chromatographic chamber saturation on the separation of L-584,428 using RP-18. Mobile phase: acetonitrile–methanol–0.1 M sodium acetate–triethylamine (61:18:7:14, v/v). (*a*) Saturated conditions; (*b*) nonsaturated conditions

1*a* and *b*. Using silica-gel as a stationary phase and a nonsaturated chromatographic chamber led to faster migration of the solute. On the other hand, saturation of the chromatographic chamber with mobile-phase vapors produced a compression of R_f values with a slight improvement in resolution between the main spot and the impurities present (not shown).

Conversely, on RP-18 (Figures 6.2*a* and *b*) the saturation of the chromatographic chamber had a negative effect. In this case, the saturated tank produced diffusion of the spots with a loss in resolution.

Influence of Silica-Gel Activation

The activity of the silica-gel layer is related to the amount of moisture present. Silica-gel contains water that has been trapped in it porous structure which can influence the mechanism of separation. Consequently, the R_f values of separated compounds may vary, depending upon the moisture content in the silica-gel layer. This water can be partially removed by heating the TLC plate. However, the readsorption of water is a rapid process and leads to a nonactivated surface (8). Maintaining plate activation requires conditions of controlled temperature and humidity (9). Establishing these conditions is costly and impractical in a laboratory environment.

To evaluate the effect of layer activation on the separation, a silica-gel HPTLC plate was heated at 120 °C for 30 min. Following this we deactivated part of the plate by dipping half of it into a solution of 20% water in acetone. In this way a TLC plate that was half activated and half deactivated was prepared. The results depicted in Figure 6.3 show no difference in selectivity between the activated part of the layer and the nonactivated portion. Displacement of adsorbed water by triethylamine present in the mobile phase could account for this phenomenon. Rapid rehydration of the activated part of the layer may also explain this observation.

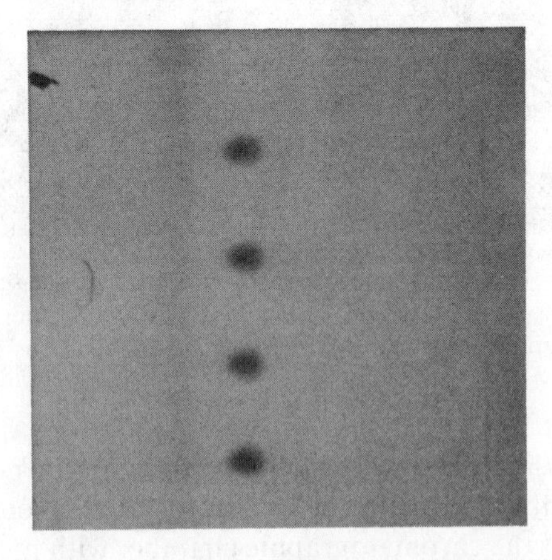

Figure 6.3. Influence of layer activation upon separation of L-584,428 on silica gel HPTLC. Conditions are the same as in Figure 6.1. The first two spots (from left to right) are on the nonactivated portion of the plate. The remaining two spots are on the activated portion.

On-Plate Decomposition

Decomposition occurring during a separation can be evidenced by applying the same kind of interaction in two directions. A slight improvement in resolution (increased by a factor of the squared root of 2) may be observed due to an increased path length traveled by the solute. In addition, a diagonal pattern of impurities should be obtained if no decomposition is occurring during the chromatographic process. Conversely, if on-plate decomposition occurs, off-diagonal spots can be obtained.

The occurrence of multiple spots in TLC has been thoroughly studied. There are several sources of multiple spots in TLC: chemical reaction (10), impurities present in the sample solution (11), discontinuities in the stationary phase (12), discontinuities in the mobile phase (13), formation of charged species and/or complexes (14), and equilibrium between species (15).

In our study of MK0912, artifacts were detected due to stationary phase catalyzed chemical reactions. If the kinetics of decomposition is fast, then well-defined spots can be seen off diagonal. Conversely, if the kinetics of decomposition is slow, then streaking will occur on the plate. Figures 6.4*a* and *b* illustrate these phenomena.

The above phenomenon was studied with different analytes. MK0912 was used as a model compound and was chromatographed on a HPTLC silica-gel plate using a mobile phase of ethyl acetate–hexane–methanol–ammonia (50:35:15:2, v/v). The analyte solution was applied at the origin of the TLC plate as a band. The plates were visualized under short-wave and long-wave UV. A photograph of the chromatogram under short-wave UV is presented in Figure 6.5*a*. A major spot was observed at an R_f of 0.43. By inspecting the

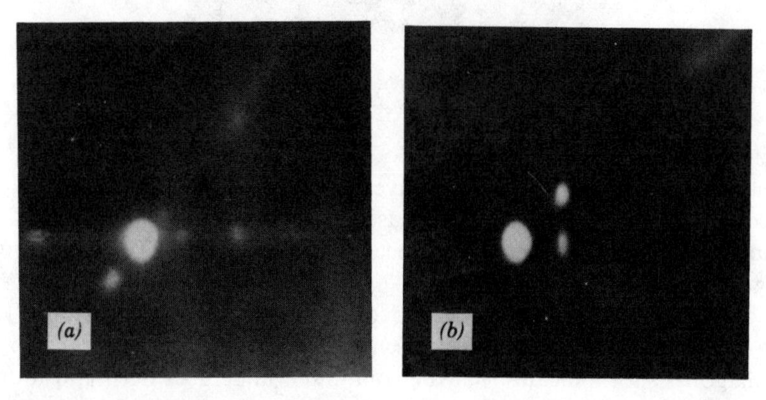

Figure 6.4. Two-dimesional TLC showing (*a*) slow kinetics of decomposition and (*b*) fast kinetics of decomposition.

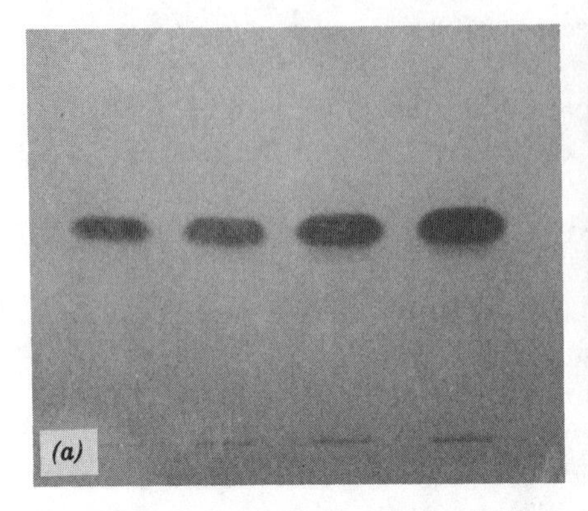

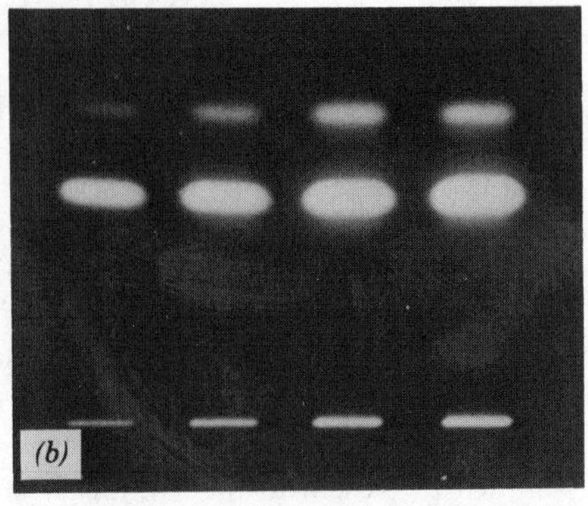

Figure 6.5. Separation of MK0912 on silica-gel HPTLC. (*a*) Visualization short-wave UV; (*b*) visualization long-wave UV. Mobile phase: ethyl acetate–hexane–methanol–ammonia (50:35:15:2, v/v). The amount spotted increases from left to right (1.25, 2.5, 5.0, and 7.5 μg).

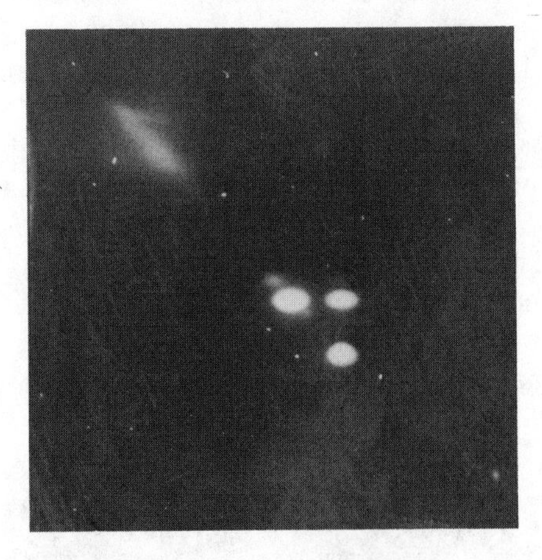

Figure 6.6. Two-dimensional TLC of MK0912. Conditions are the same as in Figure 6.5.

same chromatogram under long-wave UV, additional spots were encountered. The intensity of the spots increased after heating the plate at 120 °C for 10 min. The extraneous spots were located at R_f values of 0.0, 0.54, and 0.59 (Figure 6.5*b*).

To probe the system for on-plate decomposition, a two-dimensional TLC was performed and the result is shown in Figure 6.6. The appearance of an off-diagonal spot indicates that the compound undergoes decomposition.

A stability study was performed to establish whether the decomposition occurs in the sample solution or under chromatographic conditions (in the

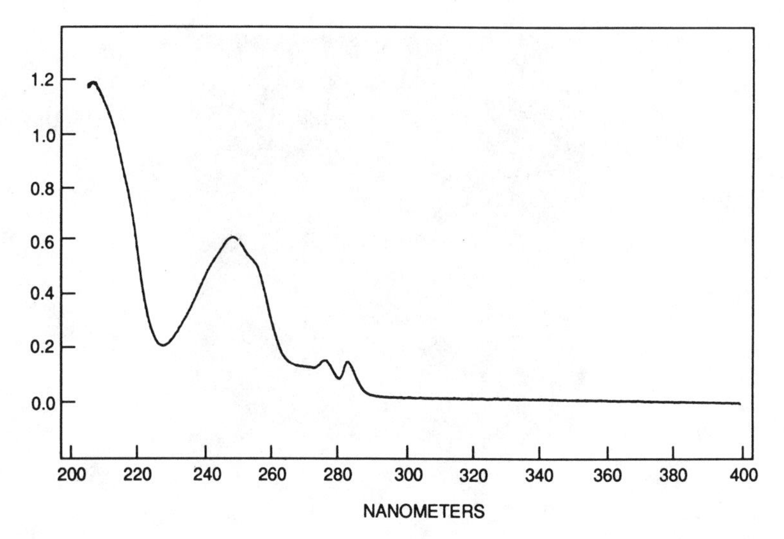

Figure 6.7. UV spectra of MK0912 during a 1–h time interval.

mobile phase or on the stationary phase). First, UV spectra of the compound dissolved in methanol were recorded at 10 min intervals for 1 h. Figure 6.7 depicts the spectra taken during the 1 h period. The spectra were perfectly superimposable, suggesting that no decomposition occurred in the time frame examined. A similar study was conducted to examine the stability of MK0912 in the mobile phase and no spectral changes were noted.

The influence of the stationary phase on the decomposition of MK0912 was then investigated. Each spot was scraped from the stationary phase and independently subjected to two-dimensional TLC. It is interesting to note that the two-dimensional TLC of each exhibited an off-diagonal pattern implying on-plate decomposition. To determine the effect of the stationary phase on the decomposition of MK0912, several spots all of equal amounts were spotted on a plate at 10 min intervals. The time elapsed between the spotting of the first and last spots was 60 min. Figure 6.8 shows the results of this study.

It is clear that the intensities of the leading spots (R_f values of 0.54 and 0.59) are less in the latter run than in the initial run. The above results indicate that the extraneous spots were generated on the plate. This suggests the stationary phase is responsible for the degradation of the compound.

There are many ways to avoid on-plate chemical decomposition. For example if the compound undergoes oxidation during the spotting procedure or chromatography, the addition of an antioxidant such as BHT (16) can overcome the problem (Figure 6.9). If decomposition occurs during the

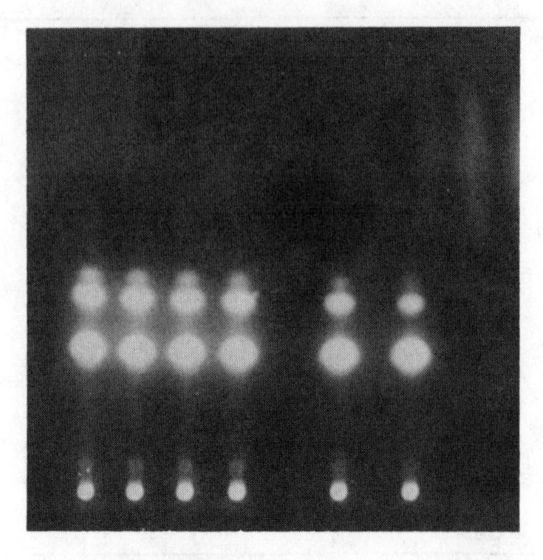

Figure 6.8. The influence of stationary phase on decomposition of MK0912.

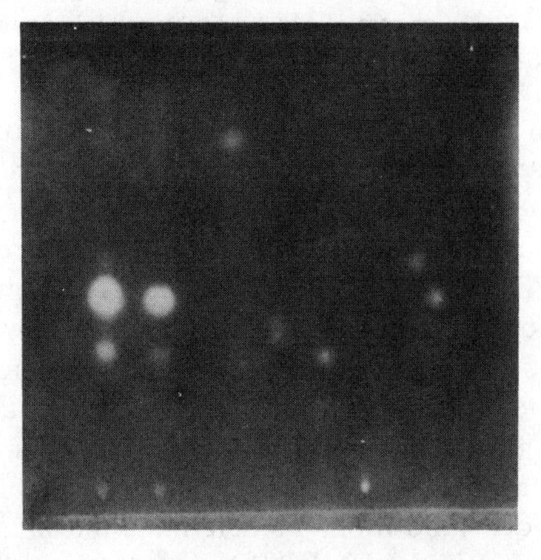

Figure 6.9. RP-18 HPTLC of L-654,969 spiked with known impurities. Mobile phase: acetonitrile–methanol–0.1 M ammonium acetate (70:10:20, v/v). Spotting solvent:0.05% BHT in methanol.

chromatographic separation, then the use of micellar chromatography may help to maintain the structural integrity of the compound (17).

CONCLUSIONS

In the present study several parameters were emphasized that are critical in the development of a TLC method for a pharmaceutical. The investigation has shown the important role of chromatographic chamber saturation. The experimental results also indicated that the activity of the silica-gel layer had no influence on chromatographic selectivity for the compounds examined. The on-plate degradation study showed the importance of correctly interpreting and characterizing the impurity profile. Since the amount of impurities present is a central issue, correct evaluation is crucial in ensuring the effectiveness of a drug candidate.

REFERENCES

1. E.L. Inman, J.K. Frischmann, P.J. Jimenez, G. D. Winkel, M. L. Persinger, B. S. Rutherford, *J. Chromatogr. Sci.* 25: 252 (1987).
2. R. A. De Zeeuw, *J. Chromatogr.* 32: 43 (1968).
3. M. S. J. Dallas, *J. Chromatogr.* 17: 267 (1965).
4. M. Brenner and A. Neiderwieser *Experienta* 16: 378 (1960).
5. E. Stahl, *Z. Anal. Chem* 221: 3 (1966).
6. T. C. Ovenston, *Nature* 169: 924 (1952).
7. Kirschner, J. M. Miller, and G. J. Keller, *Anal. Chem.* 23: 420 (1951).
8. R. P. W. Scott and S. Traiman, *J. Chromatogr.* 196: 193 (1980).
9. R. A. DeZeeuw, in *Progress in TLC and Related Methods*, Vol. 3, A. Niederwieser and G. Pataki, (Eds.). Ann Arbor Science, Ann Arbor, MI, 1972, p. 39.
10. L. Zechmeister and L. Cholnovky *Principles and Practice of Chromatography*. Wiley, New York, 1951, p. 5.
11. H. H. Strain, *Ind. Eng. Chem.* 42: 1307 (1950).
12. S. Moore and W. H. Stein, *Annu. Rev. Biochem.* 21: 521 (1952).
13. H. G. Boman, *Nature* 163: 215 (1949).
14. A. Landua, R. Fuerst, and R. A. Awapar, *Anal. Chem.* 23: 162 (1951).
15. S. M. Partridge and R. G. Westall, *Biochem. J.* 42: 238 (1948).
16. T. S. Neudoerffer and C. H. Lea *J. Chromatogr.* 21: 138 (1966).
17. N. Grinberg, Unpublished results.

CHAPTER

7

ENANTIOMERIC SEPARATION BY THIN-LAYER CHROMATOGRAPHY

SOON M. HAN AND DANIEL W. ARMSTRONG

INTRODUCTION

Enantiomeric separations are very important in many fields, and a variety of chromatographic approaches have been utilized successfully. Unfortunately, planar chromatographic methods have lagged behind their LC counterparts and there are very few reports on the separation of enantiomers. There are two methods for the separation of enantiomers by thin-layer chromatography (TLC). The first involves the use of a chiral stationary phase, while the second involves chiral mobile-phase additives.

In 1980, Yuasa and coworkers first reported separation of DL-tryptophan, DL-histidine, DL-phenylalanine, DL-tyrosine, and DL-β-3,4-dihydroxyphenylalanine using crystalline cellulose TLC plates (1). Wainer and coworkers have reported the application of γ-aminopropyl silanized silica layer dynamically coated with (*R*)-*N*-(3,5-dinitrobenzoyl)phenylglycine for the separation of racemic 2,2,2-trifluoro-1-(9-anthryl)ethanol into its isomers. (2). The plate was developed with 9.5:1 hexane:isopropyl alcohol. The two fluorescent spots had R_f values of 0.59 (−isomer) and 0.49 (+ isomer) and the separation factor (α) was 1.5.

The separation of some dansyl amino acids was performed by Weinstein on reversed-phase plates treaated with copper complex of *N*,*N*-di-*N*-propyl-L-alanine (DPA). (3). All the dansyl protein amino acids except proline were separated. This method was further developed by Grinberg and Weinstein using two-dimensional techniques (4). In the first dimension, dansyl amino acids were separated into components in a nonchiral mode by using a convex gradient elution with aqueous NaOAc buffers varying concentration of MeCN. In the second dimension, the plates were treated with the chiral copper complex of (*N*,*N*-di-*N*-propyl-L-alanine and were developed in aqueous MeCN–NaOAc buffers. To optimize the separation of each dansyl amino acid, a temperature gradient was applied.

Commercially available chiral plates were used by Brinkman and

Kamminga for the separation of enantiomers of underivatized amino acids. These chiral phases were made by treating octadecyl-modified silica plates with a solution of copper acetate followed by a solution of (2*S*,4*R*,2′*RS*)-4-hydroxy-l-(2-hydroxydodecyl)proline. Methanol:water:acetonitrile (50:50:200) was used as the eluent, while methanolic ninhydrin was used for detection (5). Subsequently, this chiral plate was used for enantiomeric separation of *N*-carbamyl tryptophan (6), methyldopa (7), and the separation of α and β anomeric forms of adenosine and deoxyadenosine (8).

Marchelli and coworkers (9) separated dansyl amino acids by one- and two-dimensional TLC with or without chiral additives in the eluent (water–acetonitrile) and under isocratic conditions. In the former, reversed-phase C_{18} plates with $RC(NH_2)HC(O)NH(CH_2)_nNHC(O)C(NH_2)HR$ ($R = Ph\ CH_2$, $n = 2$) and copper acetate additives were used at different pHs.

Cyclodextrins have been used successfully in chromatography for the separation of enantiomers, diastereomers, and structural isomers (10–15). Most of these separations have been done via high-performance liquid chromatography (HPLC) with cyclodextrin bonded phase. More recently, cyclodextrins have been shown to be particularly useful in the TLC separation of a variety of optical isomers. They have been used both as a bonded phases and as mobile-phase additives. To understand, properly, the use of cyclodextrins in TLC, it is necessary to understand the structure and properties of the cxyclodextrins in TLC. (16–18).

Cyclodextrins are cyclic oligosaccharides that contain 6 to 12 glucose units bonded through α-(1,4) linkages. Three cyclodextrins, α-cyclodextrin (cyclohexaamylose), β-cyclodextrin (cycloheptaamylose), and γ-cyclodextrin (cyclooctaamylose), are commercially available. Cyclodextrins are chiral and toroidal-shaped molecules. The inside of the cyclodextrin cavity is relatively hydrophobic and the outside of the cyclodextrin is hydrophilic, so it can complex with water insoluble or slightly soluble compounds. The internal diameter of α-cyclodextrin varies from about 4.5 to 6.0 Å which is a good size for complexing a single or six-membered aromatic ring. The internal diameter of β-cyclodextrin varies from 6.0 to 8.0 Å and can easily accomodate molecules the size of biphenyl or naphthylene. The internal diameter of γ-cyclodextrin varies from approximately 8 to 10 Å. Molecules as large as substituted pyrenes can bind to γ-cyclodextrin. Among the parameters that determine whether an inclusion complex can be formed are the relative size and geometry of the guest molecule in the host cavity, van der Waals interactions, hydrophobic interactions, hydrogen bonding, and so on. The different sized cyclodextrins allow one to separate a variety of different sized enantiomers.

EXPERIMENTAL SECTION

Cyclodextrin Bonded Phase

Materials

Seven types of silica gel were used: (1)Thorn Smith silica gel (TLC-7GF), (2) ASTEC 40-μm-diameter, 300-Å pore size silica gel, (3) ASTEC 5 to 40-μm-diameter, 60-Å pore size silica gel, (4) ASTEC 10-μm-diameter, 60-Å pore size silica gel, (5) ASTEC 3-μm-diameter, 60-Å pore size silica gel, (6) Whatman K5, 150-Å pore size silica gel, and (7) Macherey Nagel 5 to 20-μm, 60-Å pore size silica gel. The three types of binders used in this study were (1) ASTEC acrylate binder, (2) poly(ethylene glycol) (Aldrich), and (3) ASTEC "All Solvent Binder." β-Cyclodextrin was obtained from Advanced Separation technologies, Inc. (ASTEC). All dansyl amino acids and β-naphthylamide amino acids were obtained from Sigma (St. Louis, MO). Nitroaniline, nitrophenol, and stilbene isomers were obtained from Aldrich (Milwaukee, WI). *cis*- and *trans*-benzo [*a*]-pyrene-7,8-diol were the generous gift of H. J. Isaaq (Frederick Cancer Research Facility). Ferrocene enantiomers were produced as previously reported. (19). HPLC-grade methanol and water were obtained from Fischer Scienticfic (St. Louis, MO).

Methods

TLC plates (5 × 20 cm) were prepared by mixing 1.5 g of β-cyclodextrin bonded silica gel in 15mL of 50% methanol (aq) with 0.002 g of binder. The slurry was spread to a thickness of 3 mm on a clean glass plate and left to air-dry. The plate was then heated in an oven to 75 °C for 15 min before use. The stationary-phase thickness was < 1 mm on the finished plate. All developments were done at room temperature (20 °C) in an $11\frac{3}{4}$ in-long, 4-in-wide, and $10\frac{3}{4}$ in-high Chromaflex developing chamber. Spot visualization was done by use of a fixed-wavelength (254-nm) lamp.

Cyclodextrin as Chiral Mobile-Phase Additives

Materials

Chemically bonded octadecylsilane reversed-phase TLC plates, KC18F (200 μm layer thickness, 5 × 20 cm and 20 × 20 cm) were obtained from Whatman Chemical Separation Division, Inc. (Clifton, NJ). β-Cyclodextrin was obtained from Advanced Separation Technologies, Inc. (Whippany, NJ)

Table 7.1. Properties of Substituted β-Cyclodextrins[a]

Compounds	Average Molar Substitution[b]	Average Molecular Weight[c]	Solubility in H_2O (g/100 mL, 25 °C)	Other Solvents[d]
Hydroxypropyl-β-cyclodextrin	0.6+0.1	1379+40	96	MeOH, EtOH, DMF DMSO, pyridine
Hydroxypropyl-β-cyclodextrin	0.9+0.1	1501+40	>100	MeOH, EtOH, DMF DMSO, pyridine
Hydroxyethyl-β- cyclodextrin	1.0+0.2	1443+62	>200	MeOH, EtOH, DMF DMSO, pyridine
Hydroxyethyl-β-cyclodextrin	1.6+0.2	1628+62	>200	MeOH, EtOH, DMF DMSO, pyridine

[a] Data obtained from the Consortium fur Elekctrochemische Industrie GMBH technical literature.
[b] Average molar substitution is defined as the average number of hydroxypropyl or hydroxyethyl groups per anhydroglucose unit.
[c] Each sample contained a narrow distribution of homologues. For example, the molar substitution of the first compound in this table varied from 0.5 to 0.7 for an average of 0.6. Consequently, the molecular weight varied from 1338 to 1419 for an average of 1379.
[d] The solvents listed here will also dissolve the substituted cyclodextrin to some extent.

Source: Armstrong et al. (30).

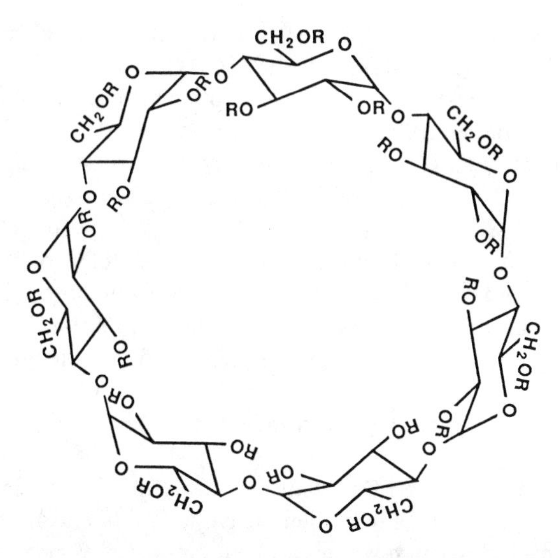

Figure 7.1. The basic structure of derivated cyclodextrin is shown. For hydroxyethyl-β-cyclodextrin, $R=(CH_2\text{–}CH_5\text{–}O)_nH$. For hydroxypropyl-β-cyclodextrin,

$$R = (CH_2\text{–}\underset{\displaystyle CH_3}{\underset{|}{CH}}\text{–}O)_nH$$

The cyclodextrin is randomly substituted and side chains with $n > 1$ are possible. the average molar substitution (AMS) of each derivative (i.e., 0.6, 0.9, 1.0, and 1.6 in Table 7.1) refers to the number of hydroxyalkyl groups per anhydroglucose unit. From Armstrong et al. (30).

and Ensuiko Sugar Refining Co. (Japan). Partially substituted hydroxypropyl-β-cyclodextrin and hydroxyethyl-β-cyclodextrin (Table 7.1) were obtained from Otto Huber of the Consortium fur Electrochemishe Industrie GMBH, Zielstattstrasse 20, 8000 Munchen 70. The basic structure of these compounds is shown in Figure 7.1. All dansyl amino acids, cinchonine, cinchonidine, quinine, quinidine, 4-pregnene-17α,20α-diol-3-one,4-pregnene-17α,20β-diol-3-one, 4-pregnene-17α,20β,21-triol-3,11-dione,4-pregnene-17α,20β,21-triol-3,11-dione,20-hydroxy 4-pregnene-3-one, 20β-hydroxy 4-pregnene-3-one, 4-pregnene-17α,20α,21-triol-3,11-dione, 4-pregnene-17α,20β, 21-triol-3, 11-dione, 4-pregnene-11β,17α,20α, 21-tetrol-3-one, 4 pregnene-11β,17α,21β,21-tetrol-3-one, and sodium chloride were obtained from Sigma (St. Louis, MO). ±2-Chloro-2-phenylacetyl chloride, DL-alanine-β-naphthylamide hydrochloride, (1*R*,2*S*,5*R*)-(-)-menthyl-(S)-*p*-toluenesulfinate, (1*S*,2*R*,5*S*)-(+)-menthyl-(*R*)-*p*-toluenesulfinate, α-ethyltryptamine, DL-methionine-β-naphthylamide, 5-(4-methylphenyl)5-phenylhydantoin, *R*-(=) benzyl-2-oxazolidinone, and *S*-(−) benzyl-2-

oxadolidinone acetate were obtained from Aldrich (Milwaukee, WI). Urea and sodium chloride were obtained from MCB (Cincinnati, OH). Ferrocene enantiomers (−)*S*-(1-ferrocenyl-2-methylpropyl)thioethanol, (+)*S*-(1-ferrocenylethy)ethiophenol), nicotine enantiomers (*N*′-benzylnornicotine, *N*′-(2-naphthylmethyl)nornicotine), *N*′-menthoxycarbonyl)anabasine, *N*′-(methoxycarbonyl-3-pyridyl-1-aminoethane, and crown ether enantioomers (+)2,2-binaphthyldiyl-*N*-benzylmonoaza-16-crown-5) were produced as previously reported (19–21). Mepthenytoin and labetalol were obtained from R. D. Armstrong of the La Jolla Cancer Research Foundation. HPLC grade water, acetonitrile, methanol, and triethylamine were obtained from Fisher Scientific (St. Louis, MO). All chemicals were used without purification.

Methods

The solubility of β-cyclodextrin in water is $1.67 \times 10-2$ M at 25 °C, and the solubility decreases significantly when organic modifiers are added. The solubility of β-cyclodextrin in water can be increased at very high pH values or by adding large amounts of urea. When the solution is saturated by urea, one can increase the solubility of β-cyclodextrin up to 0.2 M. In this study, saturated solutions of urea were used when β-cyclodextrin was used as mobile-phase additives. When synthetically modified cyclodextrins (partially substituted hydroxypropyl- and hydroxyethyl-β-cyclodextrin) were used as mobile-phase additives, additions of urea was not necessary. Hydroxypropyl- and hydroxyethyl-β-cyclodextrin are much more soluble in water and hydro-organic solvents than β-cyclodextrin; solutions exceeding 0.4 M can be made without including additives. The hydroxyethyl-β-cyclodextrin may be more soluble than hydroxyethyl-β-cyclodextrin. It appears that the solubility of the hydroxypropyl- and hydroxyethyl-β-cyclodextrin increases as the degree of substitution increases. 0.6 M NaCl also was added to the mobile phase to stabilize the binder of the reversed-phase plates. Without this salt, mobile phases containing $>50\%$ water tend to dissolve the binder of Whatman reversed-phase plates, thereby resulting in the separation of the stationary phase from the glass support during development.

It took approximately 6–8 h to completely develop a 20×20-cm and a 5×20-cm TLC plate with β-cyclodextrin mobile phase. In the case of concentrated solutions of hydroxypropyl-β-cyclodextrin and hydroxyethyl-β-cyclodextrin, the time of development increased substantially because the viscosity of derivatized β-cyclodextrin solutions increases with concentration (see Figure 7.2). All developments were done at room temperature (20 °C) in 6 (i.d.) × 23-cm cylindrical glass chambers and a $28.5 \times 9.5 \times 27.0$-cm glass chamber.

Spot visualization was done by use of a fixed-wavelength (254-nm) UV lamp. When the spots could not be determined by UV lamp, a Shimadzu

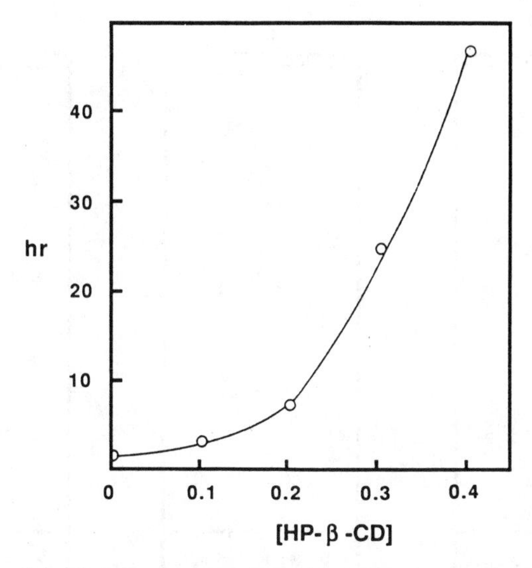

Figure 7.2. A plot of the TLC development time (in hours) versus the molar concentration of 0.6 MS hydroxypropyl-β-cyclodextrin in the mobile phase. The solute used was dansyl-D,L-leucine. The mobile phase was 30:70 (v/v) acetonitrile: water. The separation was done on a 5 × 20-cm reversed-phase TLC plate. From Armstrong et al. (30).

dual-wavelength TLC scanner (CS-910) was used to measure resolution. Single wave-length, reflection mode, and linear scanning were used. The wavelength selected correspond to that of maximum absorbance for each compound.

RESULTS AND DISCUSSION

Cyclodextrin Bonded Phase

In the production of the first effective β-cyclodextrin bonded phase TLC plate, there are three parameters that had to be optimized: the silica gel, the binder, and the mobile phase. The physical and chemical properties of silica gel control the efficiency of the separation and the bonding coverage of β-cyclodextrin. Using *o*- and *p*-nitroaniline, one can estimate the relative amount of β-cyclodextrin bonded to the silica gel. For example, *p*-nitroaniline was more strongly retained than *o*-nitroaniline if there was an appreciable amount of bonded β-cyclodextrin (see Figure 7.3). Macherey Nagel silica gel seemed to have the best coverage of β-cyclodextrin and enantiosel-

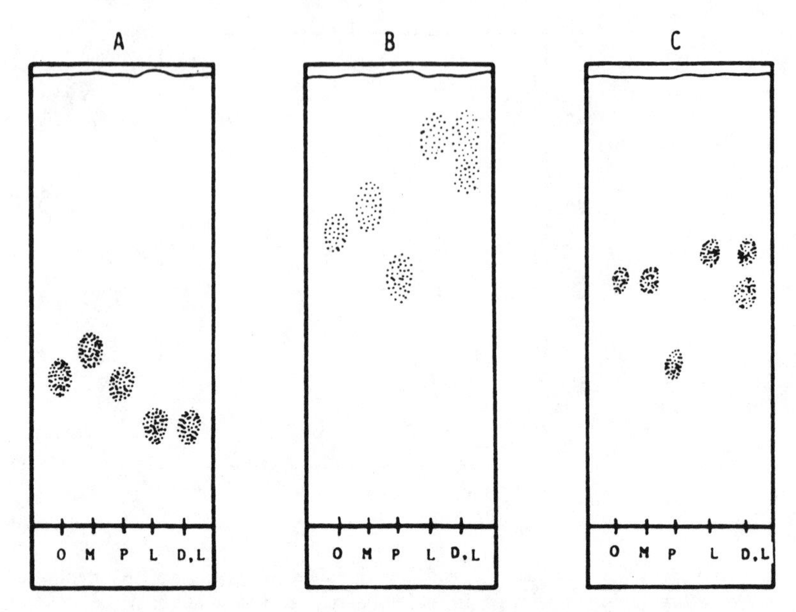

Figure 7.3. Three planar chromatograms illustrating the effect of silica gel type on the β-CD coverage and on spot size. Chromatogram A used Thorn Smith silica gel, B used Whatman, and C used Macherey Nagel silica gel (see Experimental Section for further details). L represents dansyl-L-leucine; D,L represents dansyl-D,L-leucine, O, M, and P represent *o*-, *m*-, and *p*-nitroaniline, respectively. The mobile phase was 50/50 methanol/1% aqueous triethylammonium acetate (pH 4.1). From Alak and Armstrong (22).

ectivity for dansyl-DL-leucine (22). When Thorn Smith silica gel was used, the *p*-nitoaniline retention was very similar to that of *o*-nitroaniline, and there was no separation of dansyl-DL-leucine. Each silica gel produced different efficiencies (spot size), different coverages of β-cyclodextrin, and therefore different enantioselectivities.

The binder affected the physical properties, mechanical properties, efficiency, development time, and selectivity of the TLC plates (see Figure 7.4 (22). In the case of β-cyclodextrin bonded-phase plates, one must be careful that the binder does not form a strong inclusion complex with β-cyclodextrin, thereby rendering it ineffective for further separations. Binder C (all solvent binder) showed excellent stability and mechanical properties in any solvent matrix. Binder B (an acrylate polymer) was unstable in the presence of $>50\%$ water (22).

In some cases, methanol and acetonitrile were used as organic modifiers and 1% triethylammonium acetate (TEAA), buffer, pH 4.1, was used instead of pure water. It was found that the use of buffer tended to increase the resolution and efficiency of most ionizable solutes.

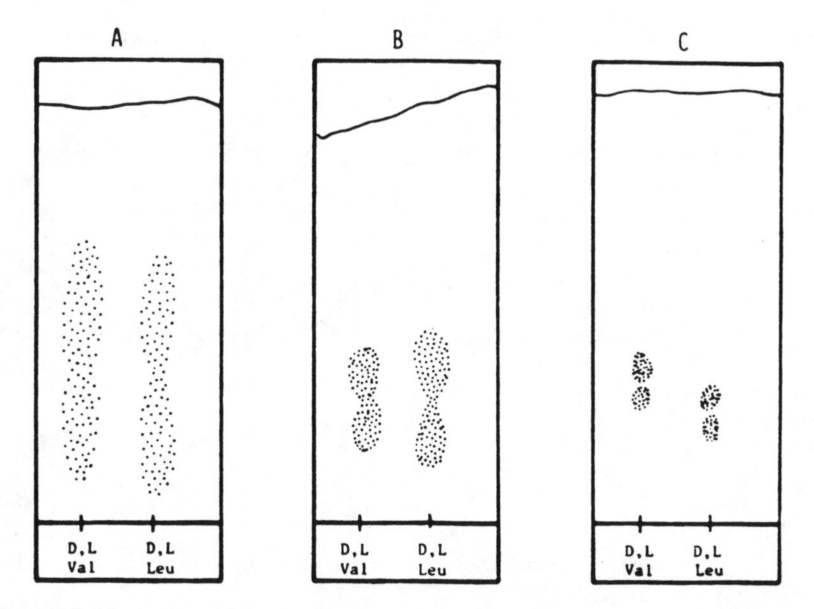

Figure 7.4. Three planar chromatograms showing the effect of binder on spot size. ASTEC 10-mm silica gel was used in all cases. Chromatogram A has no binder; chromatogram B contains an acrylate polymer; chromatogram C contains ASTEC "All Solvent Binder" (see Experimental Section for further details). The mobile phase was 50/50 methanol/1% aqueous triethylammonium acetate (pH 4.1). From Alak and Armstrong (22).

Table 7.2 shows separation data for enantiomeric compounds, diastereomers, and structural isomers using β-cyclodextrin bonded phase TLC plates. All enantiomers were resolved. The separation of DL-methionine was particularly impressive, although this mixture may have been in the sulfoxide form. In summary, β-cyclodextrin bonded phase in TLC is applicable to a wide variety of separations including enantiomers, diastereomers, structural isomers, and nonisomeric compounds.

Cyclodextrin as Chiral Mobile-Phase Additives

The first reports on the use of cyclodextrin as mobile-phase additives appeared in 1980 (23, 24). They were used to separate structural and geometric isomers on polyamide stationary phases. In 1982 cyclodextrins were used as mobile-phase additives to separate racemic mandelic acid and its derivatives (25). Since this time there have been only a few reports on the use of cyclodextrin as mobile-phase additives in LC(26–28). However, there have been no reports on the use of cyclodextrin as mobile-phase additives to

Table 7.2. Separation Data for Enantiomers, Distereomers, and Structural Isomers

Compounds	R_{f1}	R_{f2}	Mobile Phase
1. Dansyl-DL-leucine	0.49	0.66	40:60 MeOH:buffer (pH 4.1)
2. Dansyl-DL-methionine	0.28	0.43	25:75 MeOH:buffer (pH 4.1)
3. Dansyl-DL-alaine	0.25	0.33	25:75 MeOH:buffer (pH 4.1)
4. Dansyl-DL-valine	0.31	0.42	25:75 MeOH:buffer (pH 4.1)
5. DL-Alanine-β-naphthylamide	0.16	0.25	30:70 MeOH:buffer (pH 4.1)
6. DL-Methionine-β-napthylamide	0.16	0.24	30:70 MeOH:buffer (pH 4.1)
7. (±)-1-Ferrocenyl-1-methoxyethane	0.31	0.42	90:10 MeOH:buffer (pH 4.1)
8. (±)-1-Ferrocenyl-2-methylpropanol	0.33	0.39	90:10 MeOH:buffer (pH 4.1)
9. (±)-*S*-(1-Ferrocenylethyl)thioglycolic acid	0.37	0.44	90:10 MeOH:buffer (pH 4.1)
Diastereomers			
1. Quinine	0.38		25:75 MeOH:buffer (pH 4.1)
Quinidine	0.46		
2. *trans*-Stilbene	0.38		80:20 MeOH:buffer (pH 4.1)
Cis-Stilbene			
3. Benzo[*a*]pyrene-*trans*-7, 8-diol	0.46		80:20 MeOH:buffer (pH 4.1)
Benzo[*a*]pyrene-*cis*-7, 8-diol	0.52		
Structural Isomers			
1. *o*-Nitroniline	0.49		40:60 MeOH:buffer (pH 4.1)
m-Nitroaniline	0.56		
p-Nitroaniline	0.22		
2. *o*-Nitrophenol	0.56		40:60MeOH:buffer (pH 4.1)
m-Nitrophenol	0.6		
p-Nitrophenol	0.46		

Source: Alak and Armstrong (22).

separate any enantiomers by TLC. One of the reasons for this is the limited solubility of cyclodextrins (particularly β-cyclodextrin) in water and hydro-organic solvents. Enantioselectivity generally increases with increasing cyclodextrin concentration in the mobile phase. In the case of β-cyclodextrin, often the solubility limit is reached before any enantioselectivity is observed. In this study, the addition of urea permitted more concentrated β-cyclodextrin solutions which proved to be effective in the separation of enantiomers.

Figures 7.5 and 7.6 show the effect of β-cyclodextrin concentration in the mobile phase on the R_f values and resolution of dansyl-DL-glutamic acid (29). Optimum enantiomeric resolution occurs between approximately 0.08 and 0.12 M β-cyclodextrin. When the β-cyclodextrin concentration was lower than 0.02 M, there was no enantioselectivity. When β-cyclodextrin concentration was above 0.16 M, the resolution of dansyl-DL-glutamic acid

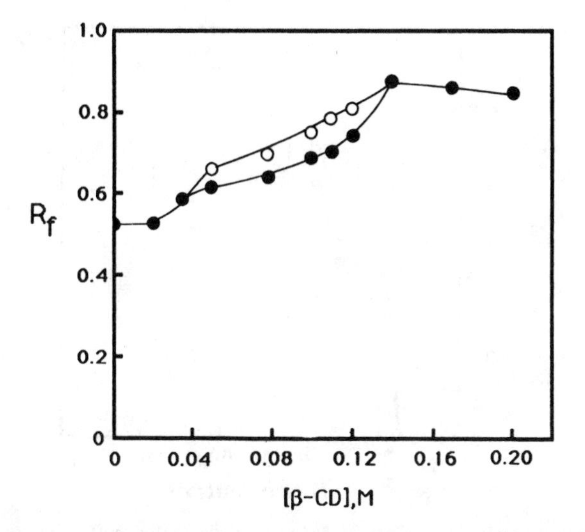

Figure 7.5. A plot showing the effect of β-cyclodextrin concentration in the mobile phase on the R_f values of dansyl-D-glutamic acid (○) and dansyl-L-glutamic acid (●). In addition to the identical levels of β-CD the mobile phase consisted of 30:70 (v/v) acetonitrile: water (saturated with urea). From Armstrong et al. (29).

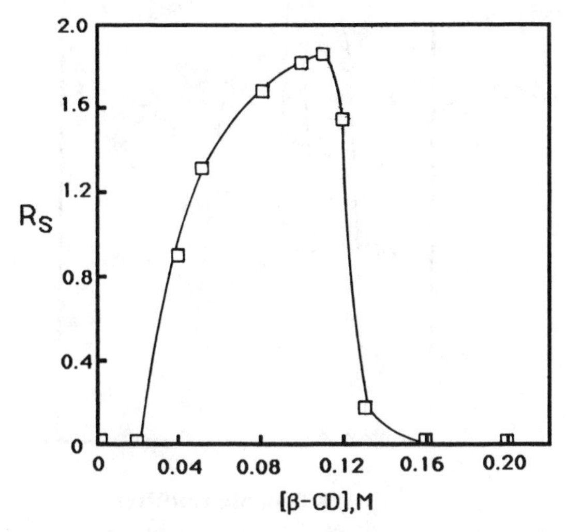

Figure 7.6. A plot showing the effect β-cyclodextrin concentration on the TLC resolution (R_s) of dansyl-D,L-glutamic acid. Other conditions are the same as in Figure 7.5. From Armstrong et al. (29).

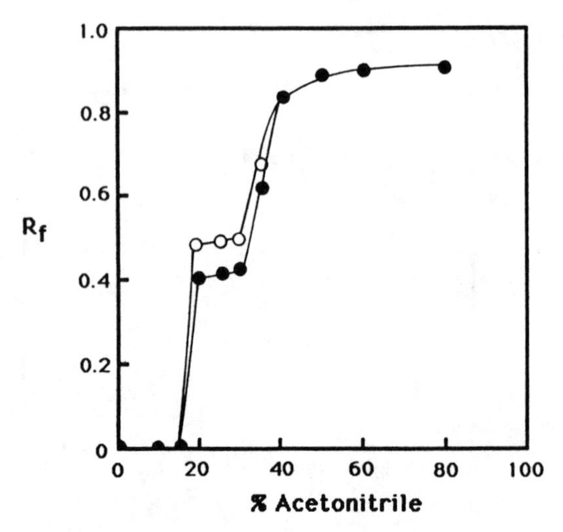

Figure 7.7. A plot showing the effect of % acetonitrile in the mobile phase on the TLC separation of dansyl-D-serine (○) from dansyl-L-serine (●). The concentration of β-cyclodextrin is 0.106M. From Armstrong et al. (29).

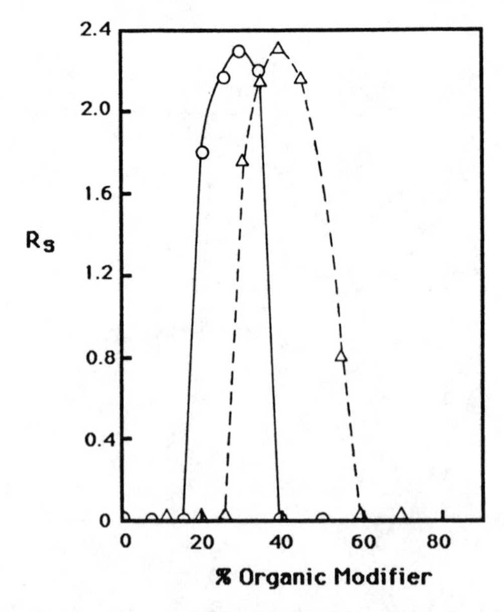

Figure 7.8. Plots showing the difference in acetonitrile (○) versus methanol (△)phase modifiers in the TLC resolution of dansyl-D,L-threonine with β-CD additives. The concentration of β-cyclodextrin is 0.106M. from Armstrong et al (29).

deteriorated as the spots blended together near the solvent front. This optimum concentration range varies with the compound and the amount of organic modifier present.

Figure 7.7 shows that the concentration and type of the organic modifier can affect enantiomeric resolution of dansyl-DL-serine. The β-cyclodextrin concentration was 0.106 M. As can be seen, enantiomeric resolution occurs over a narrow range of 10 to 15% modifier and the R_f values of solutes tend not to change appreciably in this region. When methanol was used as the modifier, the "plateau of optimum resolution" was shifted to 10% higher concentration of modifier (i.e., from 30 to 40% methanol) and to slightly higher R_f values (i.e., 0.5 to 0.6). The general effect of organic modifier type on enantiomeric resolution is shown in Figure 7.8. It is apparent that

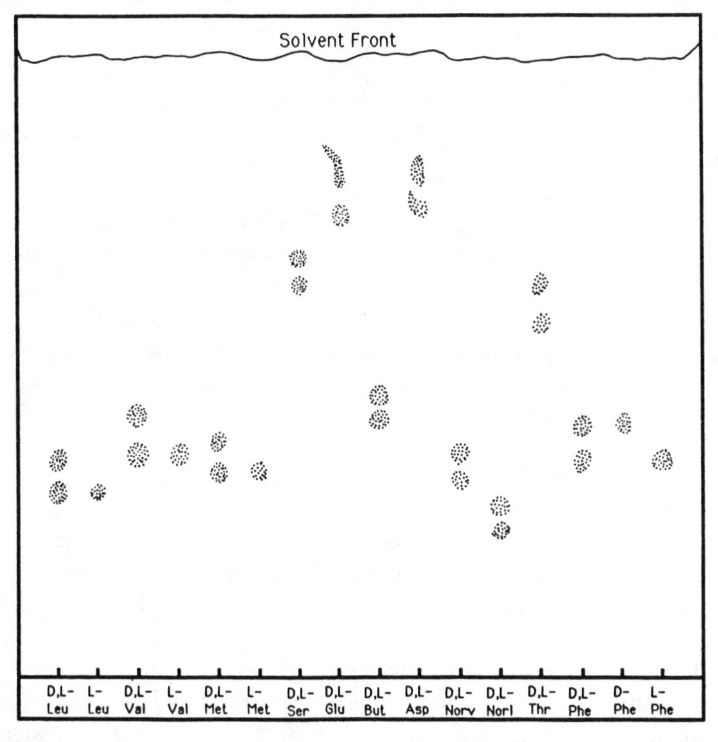

Figure 7.9. RPTLC chromatogram showing the resolution of the racemates: dansyl-D,L-leucine, dansyl-D,L-valine, dansyl-D,L-methionine, dansyl-D,L-serine, dansyl-D,L-glutamic acid, dansyl-D,L-α-amino-n-butyric acid, dansyl-D,L-aspartic acid, dansyl-D,L-norvailine, dansyl-D,L-threonine, and dansyl-D,L-phenylalaine. The mobile phase consisted of 30 : 7 (v/v) acetonitrile: 0.10 M β-CD (aq) (see Experimental Section). From Armstrong et al (29).

resolution occurs over a relatively narrow range of mobile-phase compositions, and the range for methanol is slightly greater than that for acetonitrile. The point of maximum resolution occurs at a lower % modifier in the acetonitrile case. Although Figure 7.8 shows that the resolution of dansyl-DL-threonine was the same for different organic modifiers (acetonitrile and methanol), most of the compounds show different resolution in different organic modifiers. Figure 7.9 shows the TLC chromatogram of the separation of some dansyl amino acids. As can be seen, the L-isomers elute last in all cases. This retention behaviour is opposite to that observed for the β-cyclodextrin bonded phase.

Table 7.3 lists data on the resolution of dansyl amino acids and other enantiomers. An interesting aspect of this study is that there is sometimes a poor correlation between the separation of enantiomers with the β-cyclodextrin mobile-phase additives and β-cyclodextrin bonded phases. Many compounds are separated very well by both methods. However, some enantiomers are separated by one method but not the other. For example, dansyl-DL-glutamic acid and dansyl-DL-aspartic acid are very difficult to separate on β-cyclodextrin bonded-phase LC columns and TLC plates; however, they are easily separated using cyclodextrin as a mobile-phase additive. There are a few possible reasons for this. In the bonded phase, cyclodextrin is linked to the silica gel via 8 to 10 atom spacer arms. The spacer arms can restrict the motion of the cyclodextrin and provide additional possible interaction sites for complexed molecules. Also, the surface density and configuration the cyclodextrin on the bonded-phase media may be more fixed than when using cyclodextrin as mobile-phase modifiers. Finally, the cyclodextrin is available for multiple complexation when it is used as a mobile-phase additive (26). Currently, the effect of multiple complexation and equilibria on chiral recognition is unknown. When using cyclodextrins as mobile-phase modifiers, they are both adsorbed on the achiral stationary phase and present as carriers in solution. If solution demixing occurs during development and a racemate travels ahead of the cyclodextrin solvent front, it would not be expected to resolve. Indeed this may be occurring at the higher organic modifier concentrations. However, this cannot explain why some racemates resolve via the mobile-phase additive method but not on the chiral stationary phase.

Table 7.4 lists resolution data for a variety of diastereomers. Some diastereomeric compounds are more easily separated than others. Except for cinchonine and cinchonidine, all diastereomers are baseline separated.

An alternative to the use of additives such as urea or sodium hydroxide to enhance β-cyclodextrin solubility in water or hydro-organic solvents is synthetic modification of the cyclodextrin. Figure 7.10 shows the enantiomeric resolution of dansyl-DL-leucine versus concentration of four different

Table 7.3. Separation Data for Dansyl Amino Acids and Enantiomers

Compounds	R_{f1}	R_{f2}	R_S	Mobile Phase
1. Dansyl-DL-leucine	0.30	0.35	2.0	30:70 MeCN:0.151 M β-CD
2. Dansyl-DL-valine	0.36	0.43	2.5	30:70 MeCN:0.151 M β-CD
3. Dansyl-DL-methionine	0.34	0.38	2.1	30:70 MeCN:0.151 M β-CD
4. Dansyl-DL-glutamic acid	0.65	0.72	2.0	35:65 MeOH:0.163 M β-CD
5. Dansyl-DL-α-amino-*n*-butyric acid	0.42	0.47	1.5	30:70 MeCN:0.151 M β-CD
6. Dansyl-DL-norvaline	0.32	0.34	1.4	30:70 MeCN:0.151 M β-CD
7. Dansyl-DL-norleucine	0.24	0.28	1.6	30:70 MeCN:0.151 M β-CD
8. Dansyl-DL-phenylalanine	0.35	0.39	1.4	30:70 MeCN:0.151 M β-CD
9. Dansyl-DL-serine	0.41	0.47	1.5	20:80 MeCN:0.133 M β-CD
10. Dansyl-DL-aspartic acid	0.64	0.70	1.8	25:75 MeCN:0.133 M β-CD
11. Dansyl-DL-tryptophan	0.43	0.45	0.8	35:65 MeCN:0.231 M β-CD
12. Dansyl-DL-threonine	0.42	0.51	2.0	30:70 MeOH:0.151 M β-CD
13. Mephenytoin	0.32	0.38	1.5	35:65 MeOH:0.308 M β-CD
14. Labetalol	0.49	0.53	0.7	35:65 MeOH:0.262 M β-CD
15. (±)*S*-(1-Ferrocenyl-2-methylpropyl)thioethanol	0.42	0.51	2.0	15:85 MeCN:0.125 M β-CD
16. (±)*S*-(1-Ferrocenyl ethyl)thiophenol	0.38	0.42	0.5	30:70 MeCN:0.151 M β-CD
17. *N'*-Benzyl nornicotine	0.29	0.34	1.7	60:40 MeOH:0.200 M β-CD
18. *N'*-(2-Naphthylmethyl) nornicotine	0.19	0.24	1.7	60:40 MeOH:0.200 M β-CD
19. (±)2-Chloro-2-phenyl acetyl chloride	0.02	0.07	0.55	30:70 MeCN:0.151 M β-CD
20. DL-Alanine-2-naphthyl amide hydrochloride	0.59	0.66	1.2	35:65 MeOH:0.163 M β-CD
21. (1*R*,2*S*,5*R*)-(−)-Menthyl-(*S*)-*p*-toluenesulfinate (1*S*, 2*R*, 5*S*)-(+)-Methyl-(*R*)-*p*-toluenesulfinate	0.06	0.08	0.6	30:70 MeCN:0.151 M β-CD
22. (±)-2,2′-Binaphthyldiyl-*N*-benzyl-monoaza-16-crown-5	0.05	0.08	0.6	60:40 MeOH:0.265 M β-CD

Source: Armstrong et al. (29).

hydroxypropyl-β-cyclodextrins(30). As one can see, the 0.6 average molar substituted (AMS) hydroxypropyl-β-cyclodextrin gives the best separation of dansyl-DL-leucine. There were substantial differences in the performance of the various β-cyclodextrin derivatives. It appears that as the average molar substitution of the cyclodextrin increases, the amount of the chiral mobile-phase additive needed to achieve an enantiomeric resolution also must be

Table 7.4. Separation Data for Diastereomeric Compounds

Compounds	R_f	R_S	Mobile Phase
1. Cinchonine	0.21	0.9	20:80 MeCN:0.133 M β-CD
Cinchonidine	0.18		
2. Quinine	0.19	1.5	40:60 MeOH:0.250 M β-CD
Quinidine	0.25		
3. N′-(Menthoxycarbonyl)	0.04	2.0	60:40 MeOH:0.200 M β-CD
anabasine	0.07		
4. N′-(Menthoxycarbonyl)-	0.24	3.1	60:40 MeOH:0.200 M β-CD
3-pyridyl-1-aminoethane	0.30		
5. 4-Pregnene-17α, 20α-diol-3-one	0.56	2.7	30:70 MeCN:0.151 M β-CD
4-Pregnene-17α, 20β-diol-3-one	0.48		
6. 4-Pregnene-17α,20α,21-triol-3,11-			
dione 4-Pregnene-17α,20β,21-	0.82	1.5	30:70 MeOH:0.151 M β-CD
triol-3,11-dione	0.68		
7. 20α-Hydroxy-4-pregnene-3-one	0.59	5.6	30:70 MeOH:0.151 M β-CD
20β-Hydroxy-4-pregnene-3-one	0.33		

Source: Armstrong et al. (29).

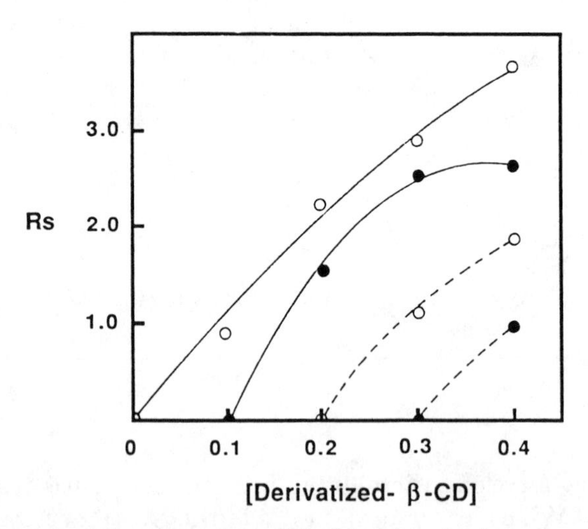

Figure 7.10. Plots of enantiomeric resolution (R_s) of dansyl-D,L-leucine versus concentration of four different hydroxyalkyl-β-cyclodextrins. The curves, from left to right are for: 0.6 AMS hydroxypropyl-β-cyclodextrin (-○),- 0.9 AMS hydroxypropyl-β- cyclodextrin (-●-), 1.0 AMS hydroxyethyl-β-cyclodextrin (--○--), and 1.6 AMS hydroxyethyl-β-cyclodextrin (--●--). From Armstrong et al. (30).

increased. A balance must be reached in which the degree of derivatization is sufficient to increase solubility and selectivity, but not high enough to interfere with complexation or chiral recognition.

Figures 7.11 and 7.12 show the R_f value and resolution versus concentration of 0.6 average molar substituted hydroxypropyl-β-cyclodextrin for dansyl-DL-leucine. The resolution of racemic dansyl leucine increases with

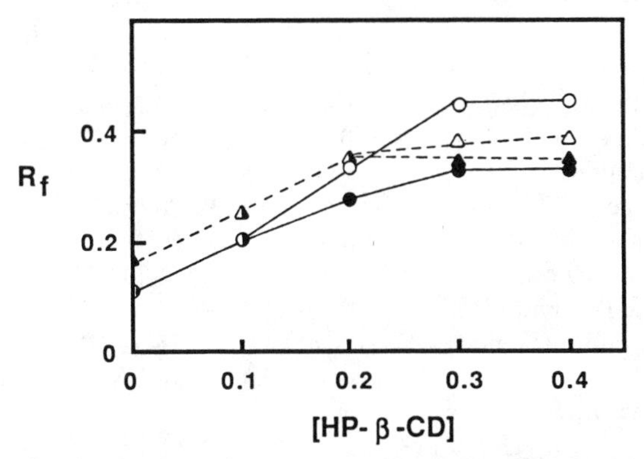

Figure 7.11. Plots showing the change in R_f versus concentration of 0.6 AMS hydroxypropyl-β-CD for dansyl-D,L-leucine (circles) and dansyl-D,L-valine (triangles). from Armstrong et al. (30).

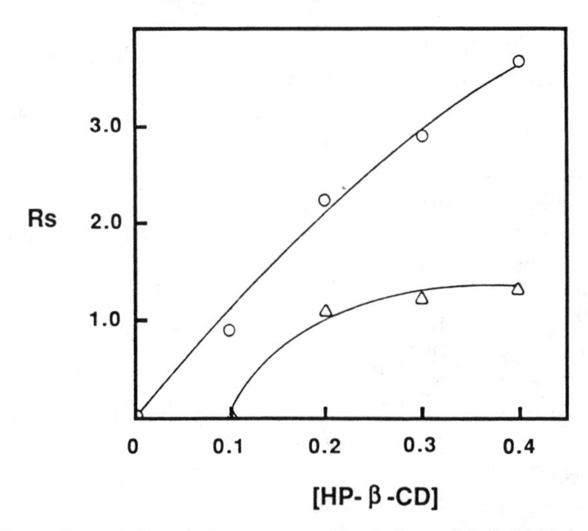

Figure 7.12. Plots of resolution (R_s) versus concentration of 0.6 AMS hydroxypropyl-β-cyclodextrin for dansyl-D,L-leucine (○) and dansyl-D,L-valine (△). From Armstrong et al. (30).

increasing cyclodextrin concentration. Table 7.5 gives separation data for enantiomers and diastereomers using 0.6 average molar substituted hydroxypropyl-β-cyclodextrin. Dansyl-DL-leucine, cinchonine, cinchonine, cinchonidine, quinine, and quinidine show better resolution using 0.6 average molar substituted hydroxypropyl-β-cyclodextrin than when using β-cyclodextrin as the mobile-phase additive. Obviously, the chromatographic selectivity of

Table 7.5. Separation Data for Enantiomers and Diastereomers

Compounds	R_f	R_S	Mobile Phase
1. Dansyl-DL-leucine	0.44	3.6	30:70 MeCN: 0.4 M HP-β-CD
2. Dansyl-DL-valine	0.37	1.3	30:70 MeCN:0.4 M HP-β-CD
3. Dansyl-DL-methionine	0.37	1.6	30:70 MeCN:0.4 M HP-β-CD
4. Dansyl-DL-threonine	0.58	0.9	30:70 MeCN:0.4 M HP-β-CD
5. Dansyl-DL-phenylalanine	0.36	1.3	35:65 MeCN:0.4 M HP-β-CD
6. Dansyl-DL-norleucine	0.20	1.8	30:70 MeCN:0.4 M HP-β-CD
7. DL-Methionine-β-naphthylamide	0.60	1.8	35:65 MeCN:0.3 M HP-β-CD
8. DL-Alanine-β-naphthylamide	0.66	1.0	25:75 MeCN:0.3 M HP-β-CD
9. Mephenytoin	0.39	2.0	30:70 MeCN:0.4 M HP-β-CD
10. 5-(4-Methylphenyl)-5-phenylhydantoin	0.24	0.8	35:65 MeCN:0.3 M HP-β-CD
11. *R*-(+)Benzyl-2-oxazolidinone S-(−)-Benzyl-2-oxazolidinone	0.54	0.9	35:65 MeCN:0.3 M HP-β-CD
Diastereomers			
1. Cinchonine	0.40	4.2	35:65 MeCN:0.3 M HP-β-CD
Cinchonidine	0.23		
2. Quinidine	0.29	4.3	35:65 MeCN:0.3 M HP-β-CD
Quinine	0.15		
3. 4-Pregnene-17α, 20α-diol-3-one	0.54	4.0	35:65 MeCN:0.3 M HP-β-CD
4-Pregnene-17α, 20β-diol-3-one	0.40		
4. 20α-Hydroxy-4-pregnen-3-one	0.37	3.3	35:65 MeCN:0.3 M HP-β-CD
20β-Hydroxy-4-pregnen-3-one	0.16		
5. 4-Pregnene-17α,20α,21-triol-3,11-dione,	0.69	2.2	30:70 MeCN:0.3 M HP-β-CD
4-Pregnene-17α,20β,21-triol-3,11-dione	0.61		
6. 4-Pregnene-11β, 17α, 20α, 21-tetrol-3-one,	0.63	0.8	35:65 MeCN:0.3 M HP-β-CD
4-Pregnene-11β,17α, 20β, 21-tetrol-3-one	0.60		
7. *N*′-(Menthoxy-carbonyl) anabasine	0.02	0.8	35:65 MeCN:0.3 M HP-β-CD
	0.04		
8. *N*′-(Menthoxy-carbonyl)-3-pyridyl-1-aminoethane	0.11	2.2	35:65 MeCN:0.3 M HP-β-CD
	0.18		

Source: Armstrong et al. (30).

β-cyclodextrin and synthetically modified β-cyclodextrin are somewhat different. It is apparent that the use of β-cyclodextrin or synthetically modified β-cyclodextrine in TLC is a very useful and powerful method for separation of enantiomers and diastereomers.

ACKNOWLEDGMENT

Support of this work by the National Institute of General Medical Sciences (BMT 1R01 GM36292) is gratefully acknowledged.

REFERENCES

1. S. Yuasa, A. Shimada, K. Kameyama, M. Yasui and K. Adzuma, *J. Chromatographic Sci.* 18: 311 (1980).
2. I.W. Wainer, C. A. Brunner, and T. D. Doyle, *J. Chromatogr.* 264: 54 (1983).
3. S. Weinstein, *Tetrahedron Lett.* 25: 985 (1984).
4. N. Grinberg and S. Weinstein, *J. Chromatogr.* 303: 251 (1984).
5. U. A. T. Brinkman and D. Kamminga, *J. Chromatogr.* 330: 375 (1985).
6. L. K. Gont. and S. K. Neuendorf, *J. Chromatogr*, 391: 343 (1987).
7. J. Martens, K. Guenther, and M. Schickedanz, *Arch. Pharm* 319: 572 (1986).
8. R. S. Feldberg and L. M. Reppucci, *J. Chromatogr.* 410: 226 (1987).
9. R. Marchelli, R. Virili, E. Armani and A. Dossena, *J. Chromatogr*, 355: 354 91986).
10. D. W. Armstrong and W. DeMond, *J. Chromatogr. Sci* 22: 411 (1984).
11. D. W. Armstrong, T. J. Ward, R. D. Armstrong and T. E. Beesley, *Science* 232: 1132 (1986).
12. D. W. Armstrong, *J. Liq. Chromatogr* 7: 353 (1984).
13. D.W. Armstrong, S. M. Han, and Y. I. Han, *Anal. Biochem* 167: 261 (1987).
14. D. W. Armstrong, S. F. Yang, S. M. Han and R. Menges, *Anal. Chem.* 59: 2594 (1987).
15. D. W. Armstrong, W. DeMond, A. Alak., W. L. Hinze, T. E. Riehl and K. H. Bui, *Anal. Chem.* 57: 234 (1985).
16. M. L Bender and M. Komiyama, *Cyclodextrin.* Springer-Verlag, Berlin, 1978.
17. J. Szejtli, *Cyclodextrins and Their Inclusion Complexes* Akademiai Kiado, Budapest, 1982.
18. W. L. Hinze in *Separation and Purification Methods*, Vol. 10, C. J. Van Oss (Ed.). Dekker, New York, 1981, p. 159.
19. D. W. Armstrong, W. DeMond, and B. P. Czech *Anal. Chem.* 57: 481 (1985).
20. J. I. Seeman, H. V. Secor, D. W. Armstrong, K. D. Timmons and T. J. Ward *Anal. Chem.*, in press.

21. D. W. Armstrong, T. J. Ward, A. Czech, B. P. Czech and R. A. Bartsch, *J. Org. Chem.* 50: 5556 (1985).
22. A. Alak and D. W. Armstrong *Anal. Chem.* 58: 582 (1986).
23. D. W. Armstrong, *J. Liq. Chromatogr.* 3: 895 (1980).
24. W. L. Hinze and D. W. Armstrong *Anal Lett.* 13: 1093 (1980).
25. J. Debowski, D. Sybilskaa, J. Jurczak, *J. Chromatogr.* 237: 303 91982)
26. D. W. Armstrong, F. Nome, L. A. Spino, T. D. Golden, *Anal. Chem.* 108: 1418 (1986).
27. D. Sybilska, J. Zukowski and J. Bojarski, *J. Liq. Chromatogr.* 9: 591 (1986).
28. J. Debowski, D. Sybilska and J. Jurczak, *J. Chromatogr.* 282: 83 (1983).
29. D. W. Armstrong, F. He, and S. M. Han, *J. Chromatogr.*in press.
30. D. W. Armstrong, J. R. Faulkner, and S. M. Han, *J. Chromatogr.*, in press.

CHAPTER

8

ASSAY OF BIPHENYL METABOLITES BY HPTLC-SPECTRODENSITOMETRY

SIDNEY S. LEVIN, JOSEPH C. TOUCHSTONE, AND DAVID Y. COOPER

A major function of the liver is the detoxification or metabolism of ingested aromatic xenobiotics. The principal enzymatic pathway for hepatic metabolism in the endoplasmic reticulum is through the mixed-function cytochrome oxidase P-450 system. Cytochrome P-450 is the term given to the group of hemoproteins that catalize mixed-function oxidase systems. These hemoproteins are the terminal oxidases in microsomal electron transport chains (1). The cytochrome oxidases act on endogenous as well as on ingested compounds. They are found in microsomal and mitochondral fractions in mammalian tissue preparations. They are pigmented and have been named through their unique characteristic absorption band.

Previous studies (2, 3) have indicated that the cytochrome oxidase system may be activated or induced by the administration of various xenobiotics prior to the pending metabolic studies. The induction affects the cytochrome activity whether at the animal, organ, tissue, cellular, or subcellular level. There are many forms of the system depending on the type of inducing agent; for example, carcinogens such as benzpyrene and methyl cholanthrene induce the P-448 form, while the P-450 form is induced by phenobarbital and other barbiturates.

Induction of the enzyme system is useful in determining the viability and increasing the activity of the preparation being studied. Biphenyl is a useful substrate for the in vitro study of the effect of xenobiotic induction (4). In the uninduced rat the major metabolite of biphenyl in the liver is 4-hydroxybiphenyl (4-OH-BP) with very small amounts of the 2-hydroxybiphenyl (2-OH–BP) and 3-hydroxybiphenyl (3-OH-BP). These are found in the bile and circulating blood plasma. However, if the rat is pretreated with phenobarbital (PB) there is a dramatic increase in the secretion of 3-OH-BP. When 3-methylcholanthrene is used as the inducing agent 2-OH-BP and 4-OH-BP levels are elevated. This difference in hydroxylation activity could be used as an indicator of the type of cytochrome oxidase inducing the enzyme system and perhaps the carcinogenicity of the inducing agent (4,5).

The method involving thin-layer chromatography (TLC) as described here is sensitive enough to use for biological samples such as bile, urine, or feces without derivitization and with only little concentration or extraction. Enzyme hydrolysis for the glucuronide or sulfate conjugates is the only sample preparation. The method gives improved separation of the 3-OH-BP and the 4-OH-BP and with the use of a scanning spectrodensitometer permits the characterization of the compounds directly on the TLC plate, not only by their R_f values but also by their absorption spectra in the ultraviolet region.

MATERIALS AND METHODS

The biphenyl used for reference and as substrate for the animal metabolic studies was purchased from Sigma Chemical Co. (St. Louis MO) The mono- and di-hydroxy biphenyls were purchased from Pfaltz and Bauer Inc. (Waterbury, CT) The solvents used were toluene, methanol and chloroform; all were glass-distilled using Omnisolve from E.M Science (Cherry Hill NJ). Hard-layer 20 × 20-cm HPTLC-HLF plates with a 150-μm-thick layer (Analtech, Newark, DE) were used for all chromatography. The plates were scored into 1-cm lanes with a Schoeffel scoring device. The standards were all prepared in methanol at a 0.1-mg/mL concentration. The β-glucuronidase used for hydrolysis of the conjugates was obtained from Sigma.

Bile, urine, and feces samples were collected from Sprague–Dawley rats that had been previously prepared with a bile fistula as described by Levin et al. (6). Some of the animals were induced with phenobarbital. The biphenyl was dissolved in ethanol (1 mg/mL) and injected intraperitonealy. Bile samples were collected at 30-min intervals for the first 2 h following injection, then at hourly intervals for the next 4 h. A 24-h bile sample was obtained the next morning and thereafter samples were collected at 24-h intervals. Urine was collected as available and combined to correspond to the bile collections. Feces were collected in 24-h units. The samples were lyophilized on a Labconco freeze-dryer. The dry powder was extracted twice with methanol and the extracts were combined. This was followed by two extractions with water which were also combined. The methanol extract was evaporated to dryness, and the dry powder was then suspended in 0.5 mL 0.1 M sodium acetate buffer (pH 5.0). The water extract was reduced to original sample volume under N_2 at 37 °C and was then adjusted to pH 5.0 with 0.1 M sodium acetate. Both extracts were hydrolized by incubating at 37 °C overnight with 600 Fishman units of β-glucuronidase.

The samples and standards were streaked across the lane of the Analtech

HPTLC-HLF layer as previously described by Touchstone et al. (7). The starting line containing the samples was then dried by heating the underside of the plate with a hair dryer and the upper surface with an infrared heat bulb for 20 min. The plates were cooled to ambient temperature and then developed in a mobile phase of toluene–methanol–chloroform (75:15:10). The solvent front was allowed to travel to 15 mm from the top of the plate. After drying, the chromatograms were scanned in reflectance mode with a Shimadzu CS9000 spectrodensitometer using a deuterium source with the monochrometer set at 250 nm and the emission or cutoff filter at 400 nm. All printout recordings were at a 1:1 basis and enhanced to give maximum peaks. As each line was scanned, the position of maximum fluorescent quenching was recorded, then the detector was positioned to obtain the UV spectrum of the compound directly on the layer. The UV spectrum from 200–400 nm was determined. The spectra were compared with Sadtler UV spectra. A series of chromatograms were developed in the same mobile phase but at 4 °C in a cold room. A clean glass plate was placed over a thin-layer plate to which samples had been applied, then both were wrapped in aluminum foil and placed in the cold room. The mobile phase and a dry chromatography tank covered and sealed with parafilm were also placed in the cold room. All were allowed to equilibrate at 4 °C for 2 h. The mobile phase was then added to the tank and was allowed to equilibrate for 20 min. The aluminum foil was removed from the TLC which was quickly placed into the tank. After development the plate was removed from the tank and after drying scanning was performed in the same manner as for the other plates.

RESULTS AND DISCUSSION

One advantage of thin-layer chromatography is that the many samples collected from serial studies can be assayed simultaneously and compared with anthentic standards. Another is that sample preparation and cleanup is usually unnecessary since the samples can be extracted directly on the plate. Biphenyl metabolism takes place in two phases which occur in rapid succession. The phase I reaction involves hydroxylation of the biphenyl, followed by phase II which is the conjugation of the hydroxyl group with the sulfate or the glucuronic acid (5,8). Therefore, it is necessary to hydrolyze the samples before analysis. Direct assay of the conjugates has not yet been developed.

Previously published methods for the separation of biphenyl and biphenyl metabolites in both HPLC and TLC methods (4,9) have indicated some difficulty in the separation of the 3-OH-BP and the 4-OH-BP. The thin-layer method described here separates the major metabolites of biphenyl at ambient temperature as can be seen by the R_f values in Table 8.1. Although the

Table 8.1. Effect of Temperature on Separation of Biphenyls on TLC

Compound	$R_f \times 100$ Ambient Temp.	$R_f \times 100$ 4 °C
Biphenyl	96.5	94.4
2-Hydroxybiphenyl	75.9	47.0
3-Hydroxybiphenyl	59.2	33.9
4-Hydroxybiphenyl	56.0	29.1
2,2′-Dihydroxybiphenyl	48.8	20.8
4,4′-Dihydroxybiphenyl[a]	28.6	9.6

[a] 4,4′-OH-BP has two other less intense fluoresence quenching spots, $R_f = 19.2$ and 63.0 at ambient temperature; see Figures 8.4*B* and *C*.

Table 8.2. Effect of Medium on Spectral Characteristics of Biphenyl

Compound	Sadtler UV Catalog: Methanol λ_{max}(nm)	TLC: Silica Gel R_f	TLC: Silica Gel λ_{max}(nm)
Biphenyl (BP)	246.5	96.5	248
2-Hydroxybiphenyl (2-OH-BP)	245, 287	75.9	242; 278
3-Hydroxybiphenyl (3-OH-BP)	250	59.2	249; 288
4-Hydroxybiphenyh (4-OH-BP)	260.5	56.0	257
2,2′-Dihydroxybiphenyl (2,2′-OH-BP)	206.5; 242; 284	48.8	212:241:277
4,4′-Dihydroxybiphenyl (4,4′-OH-BP)	261.5	(A) 28.6	268
		(B) 19.2	273
		(C) 63.0	264

R_f values of 3-OH-BP and 4-OH-BP indicate that there is adequate separation, the resolution, especially at higher concentrations, is not enough to separate the peaks in the chromatographic scanning. Lowering the temperature of the chromatography tank, the mobile phase, and the plate prior to and during the development changes some of the R_f values. There is little effect on the R_f value of biphenyl near the solvent front. The more polar the compound, the greater the change in R_f with the reduction in temperature; the resolution is increased.

Further identification of biphenyl and its metabolites was made directly on the chromatogram by scanning the individual spots *in situ* through 200–400 nm. Table 8.2 lists the absorption maxima for the biphenyl standards as found in the Sadtler Laboratories Catalog of UV spectra. The compounds listed were all dissolved in methanol. The second column in Table 8.2 lists the absorption maximum of authentic standards that were separated and scanned directly on the TLC. The thin-layer plates used in this study were fluorescent; therefore, the spots could be located by scanning at 254 nm without destructive visualization procedures. This did not affect the spectral configuration of the individual compounds, which, as seen in Figures 8.1–8.3 and 8.7, are similar to those seen in the Sadtler UV cataloge. The UV spectrum of the plate fluoresecence was scanned first and then stored as background (BK) (Figure 8.1*D*), which was subtracted from each of the subsequent UV scans of the individual samples. Touchstone and Levin previously showed that the effect of the media can cause shifts in maxima of absorption (10). Despite the fact that the absorption maximum is shifted slightly, the configuration of the spectra are similar. The scan in Figure 8.2*F* is that of 3-OH-BP, which showed maxima at 249 and 288 nm. In Figure 8.6*F* (see p. 109), which is also a scan of 3-OH-BP on another TLC the peaks are

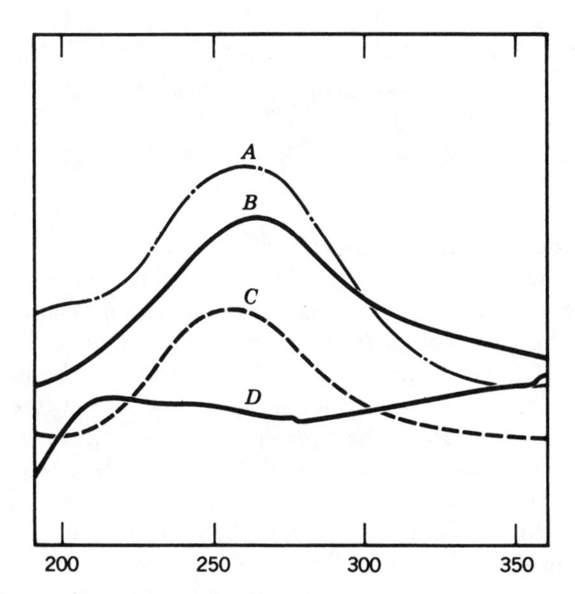

Figure 8.1 (*A–C*) UV spectra 4,4′-OH-BP: (A) most intense R_f 28.6; (B) weak adsorption R_f 19.2; (C) weak adsorption R_f 63.0. (D) UV spectral scan of Analtech HPTLC HLF fluorescent indicator. Subtracted as background from all spectra.

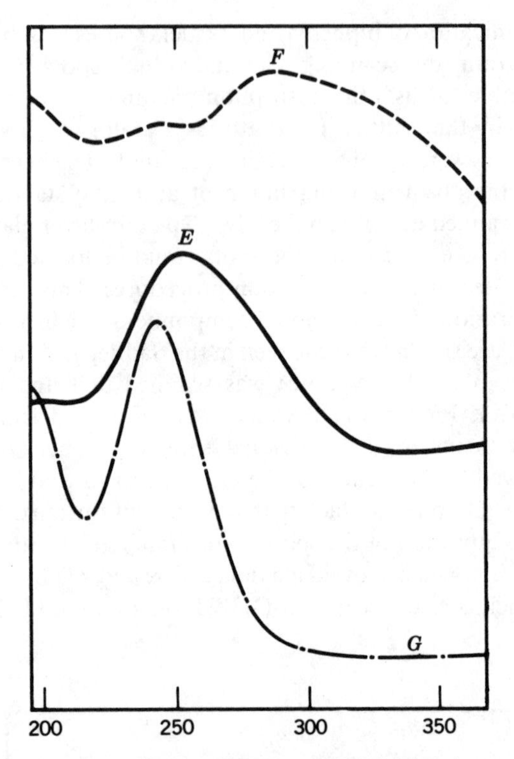

Figure 8.2. (E) UV spectrum: 4-OH-BP on HPTLC HLF plate;
(F) UV spsectrum: 3-OH-BP on HPTLC HLF plate;
(G) UV spectrum: biphenyl on HPTLC HLF plate after TLC as described.

the same but the configuration of the scan is different. This may be due to the fact that the former scan was traced from a TLC that was over a week old while the later scan was traced the same day the chromatogram was developed. The configuration of the spectra in Figure 8.6*F* more closely resembles the Sadtler spectra. The scans of Figures 8.2*E* and 8.6*E* are both of 4-OH-BP but also differ slightly in configuration. This may be due to the fact that Figure 8.2*E* was scanned with the same time delay lapse as Figure 8.2*F* while Figures 8.6*E* and *F* were scanned the same day the chromatograms were developed. Biphenyl (Figure 8.2*G*) presented problems in both spotting and storage on the chromatogram. It could not be applied to any thin-layer plate with a preadsorbent or concentrating zone; it would disappear as it moved into the active sorbent layer. A hot gun or hair dryer cannot be used to dry the samples since the biphenyl would diffuse then disappear from the plate. All spots, therefore, were dried with an infrared heat lamp placed above the

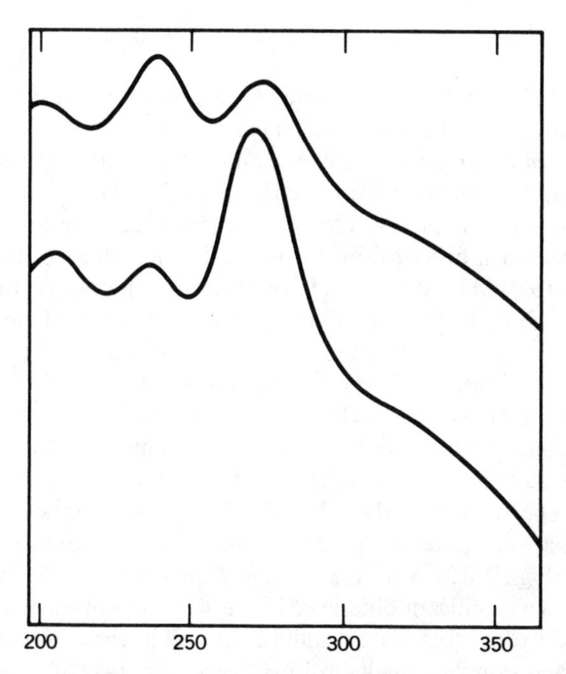

Figure 8.3. *Upper scan* : UV spectrum 2-OH-BP on HPTLC HLF plate. *Lower scan* : UV spectrum 2,2′-OH-BP on HPTLC HLF plate after TLC separation as described.

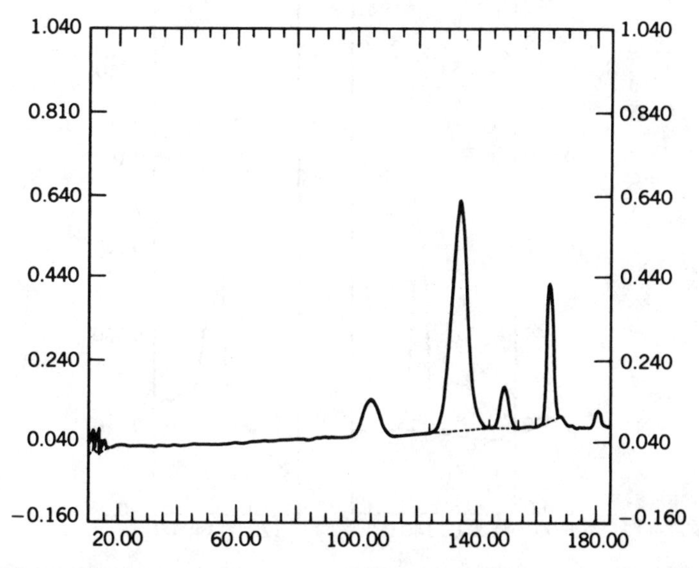

Figure 8.4. Scan of thin-layer chromatogram of a mixture of biphenyls applied to an Analtech HPTLC HLF plate; developed in toluene:methanol:chloroform (75:15:10). From left to right the peaks are 2-OH-BP; 3-OH-BP and 4-OH-BP; 2,2′-OH-BP; 4,4′-OH-BP; origin.

plate and a heat gun below the plate. A glass cover was used to protect the plate until the spectral scans were taken.

A mixture of hydroxylated biphenyl and biphenyl derivatives was applied, developed, and scanned on the same day (Figure 8.4). As can be seen in the figure, there are four peaks. The most intense peak appears as two overlapping spots when observed on the plate. The R_f values of these two spots correspond to 3-OH-BP and 4-OH-BP. In anticipation of the incomplete separation, samples of the two compounds were spotted on the adjacent lanes on either side of the mixture. These lanes were scanned (Figure 8.5), the positions of the respective spots were noted, then the UV spectra were scanned 1 mm above and below the overlapped area. The UV spectra obtained (Figures 8.6*E* and *F*) have maxima similar to the spectra of the individual standards (Figures 8.2*E* and *F*).

The UV spectra for 2-OH-BP and 2,2′-dihydroxybiphenyl (2,2′OH-BP) may at first appear to be similar except for the additional peak for the latter compound (Figure 8.3), however there is a slight shift of the maxima (Table 8.2) as well as a significant difference in the R_f values of the two compounds.

The 4,4′-dihydroxybiphenyl standards as well as the components found in bile and urine samples break up into five peaks, two of which are barely

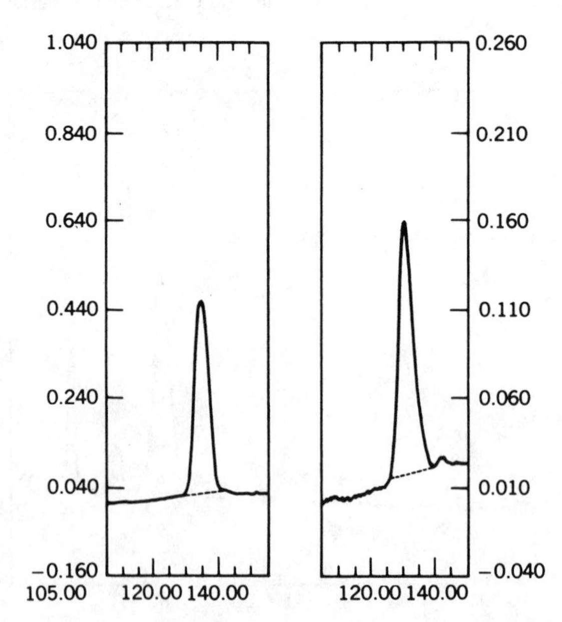

Figure 8.5. Chromatograms of authentic standards developed on an Analtech HPTLC HLF plate in toluene: methanol:chloroform (75:15:10). *Left:* 4-OH-BP PK 135; *Right:* 3-OH-BP PK 130.

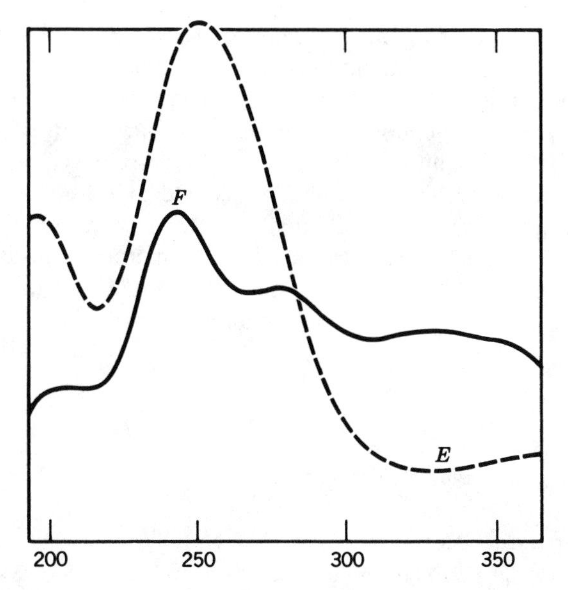

Figure 8.6. UV spectrum of chromatogram in Figure 8.5, second peak: E=4-OH-BP, F = 3-OH-BP

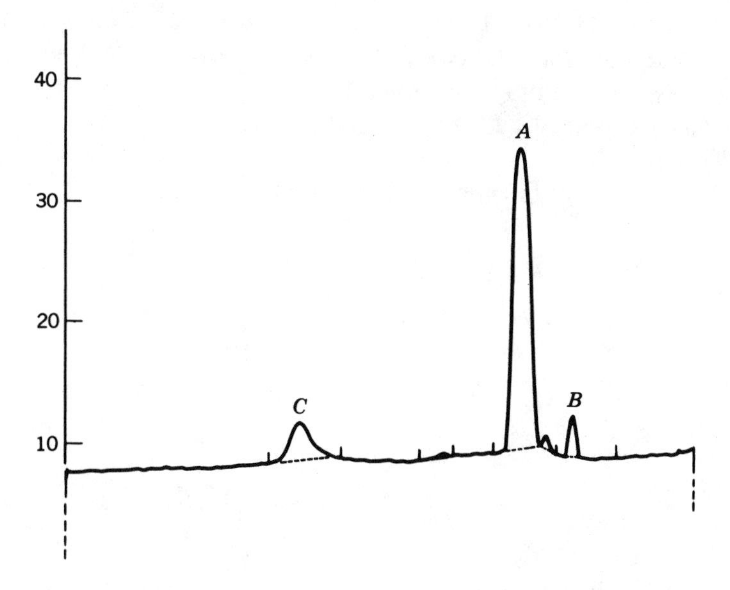

Figure 8.7. Scan of 4,4′-OH-BP standard applied to an Analtech HPTLC HLF plate and, developed in toluene: methanol: chloroform (75:15:10). (*A*) R_f 26.2; (*B*) R_f 19.2: (*C*) R_f 63.0.

visible (Figure 8.4). One of the spots has an R_f value very close to that of 4-OH-BP; however, its UV maximum (Figure 8.1*C*) is different and therefore it can be differentiated. However, it is a less intense spot and could be noted only in the presence of the most intense spot which has an R_f of 28.2 (Figure 8.7*A*) and a UV maximum of 268 nm.

The use of the biphenyl substrate administered to a Sprague–Dawley rat that has been provided with a bile fistula and has been preinduced with a suspect xenobiotic may offer a rapid screening method for suspected carcinogenic compounds. The R_f value and the in situ UV spectra provide a rapid nondestructive method of positive identification of the compounds.

REFERENCES

1. D.Y.Cooper, S.S.Levin, S. Narasimhulu, O. Rosenthal, and R.W. Estabrooke, *Science* 147: 400 (1965).
2. H.Remmer and H.J. Merker, *Science* 142: 1657 (1963).
3. E.H. Reflory and F.J. Mannering, *Mol. Pharmacol* 231: 798 (1983).
4. M.D.Burke and R.A.Prough, in *Methods in Enzymology* Vol. 52, *Biomembranes,* Fleischer and Packer (Eds). Academic, New York, 1978, p.399.
5. P. Weibkin, J.R. Fry, C.A. Jones, R.K. Lowing, and J.W. Bridges, *Biochem. Pharmacol.* 27: 1899 (1978).
6. S.S. Levin, H. M. Vars, H. Schleyer, and D.Y. Cooper, *Xenobiotica* 213 (1986).
7. J.C. Touchstone and S.S. Levin, *J. Liquid Chromotogr.* 3: 1853 (1980).
8. P. Paterson and J.R. Fry, *Xenobiotica* 15: 493 (1985).
9. G. Powis, D.J. Moore, T.J. Wilke, and K.S. Santone, *Anal. Biochem.* 167: 191, (1987).
10. J. C. Touchstone, S.S. Levin, and T. Murawec, *Anal. Chem.* 43: 858 (1971).

CHAPTER

9

IN SITU DETERMINATION OF MALONDIALDEHYDE ON THIN–LAYER PLATES BY FLUORESCENCE SPECTRODENSITOMETRY

JUAN G. ALVAREZ, BAYARD T. STOREY, AND
JOSEPH C. TOUCHSTONE

A sensitive assay for determination of malondialdehyde, produced by lipid peroxidation during aerobic incubation of tissue homogenates, is described. The method involves extraction of malondialdehyde from the homogenate, resolution by HPTLC, and in situ reaction with a 0.4% solution of TBA in 8% TCA. Scanning of the produced bands by fluorescence spectrodensitometry, using 532 nm for excitation, resulted in a lower limit of detection of 300 pg (4.2 pmol) with linear detector response extended to 2 ng (28 pmol).

The thiobarbituric acid (TBA) assay constitutes one of the most widely used methods for detection of secondary products of lipid peroxidation due to oxidation of unsaturated fatty acid moieties. One of these secondary products is malondialdehyde, a 3-carbon dialdehyde generated by homolytic clevage of cyclic five-membered ring peroxides (1) (Figure 9.1). Reaction of TBA with malondialdehyde to produce the characteristic chromogen was first observed by Kohn and Liversedge as the result of aerobic incubation of tissue homogenates (2). The chromogen was identified three years later by Bernheim et al. as the reaction product of an oxidized phospholipid (3). They showed that it was produced by the condensation of two molecules of TBA with one of a 3-carbon compound, presumably an aldehyde. Patton and Kurtz obtained a similar chromogen by reaction of TBA with malondialdehyde (4). Finally, Sinnhuber et al. isolated the TBA–malondialdehyde complex (5). Since then,the TBA assay has been extensively used to monitor lipid peroxidation in biological systems, using absorbance and fluorescence spectroscopy for quantitation (6). However, TBA has also been shown to react with other compounds normally present in these biological systems, including 2-substituted aminopyrimidines (7), dicarboxylic acids (7), sugars (8), and aminoacids (8). This lack of specificity of the TBA for malondialdehyde has resulted in the adoption of TBA reactive material as a more accurate

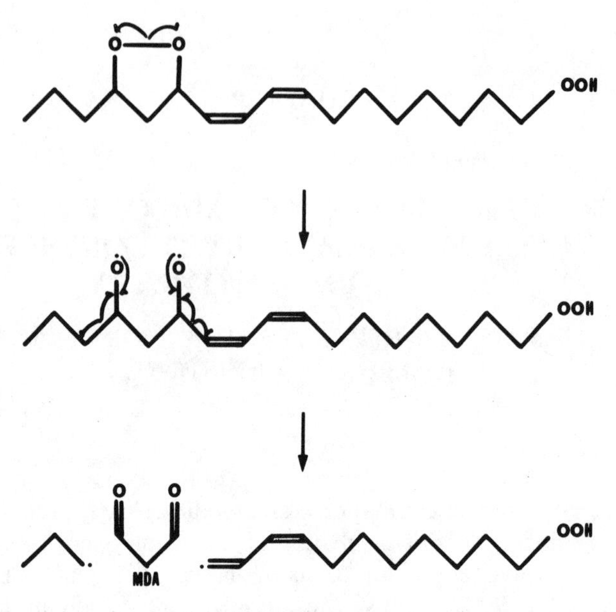

Figure 9.1. Formation of malondialdehyde (MDA) from five-membered ring cyclic peroxides.

denomination of the reation product. This prompted us to develop a method whereby reaction of the TBA with malondialdehyde, generated during aerobic incubation of tissue homogenates, could be isolated from the other components present in the system. In this paper, we present a new method of determination of malondialdehyde by in situ reaction with TBA on high-performance thin–layer plates (HPTLC).

MATERIALS AND METHODS

Reagents

Thiobarbituric acid (TBA), 1,1,3,3-tetramethoxypropane (TMP), cholesterol, and *N,N'*-ethylenediamino tetraacetic acid, disodium salt (EDTA) were obtained from Sigma Chemical Co. (St. Louis, MO). Trichloroacetic acid (TCA) and inorganic salts were from J.T. Baker (Phillisburg, NJ). Solvents were EM Science chromatographic grade. Precoated silica gel HP–K high-performance plates (250 μm thick) were obtained from Whatman Inc. (Clifton, NJ). The TBA reagent was prepared by mixing 15 mL of a 2% (w/v; 0.135 M) aqueous solution of TBA, to which enough NaOH was added for the solution to become clear, with 60 mL of 10% TCA to give a final concentration of 0.4% (w/v; 0.027 M). The TBA reagent could be kept at 4 °C for as long as one week without deterioration.

Tissue Homogenate Preparation

Rabbit testes, obtained from mature male New Zealand white rabbits, were homogenized in 50 mL of 1.2% aqueous KCl containing 0.4 μm/EDTA, by using an Ultra Turrax polytron homogenizer. No attempt was made to fractionate mitochondria and microsomes from the total homogenate. Following homogenization of the testes, 1-mL aliquots were incubated aerobically at 37 °C in a shaking water bath for 24 h. After incubation was terminated 3 mL of a solvent mixture containing chloroform–methanol–ethyl acetate (1:2:1, v/v/v) was added to 1 mL of the tissue homogenate suspension. Following centrifugation at 600g for 5 min the chloroform–methanol layer was aspirated, dried under N_2, and reconstituted in chloroform–methanol (1:1).

Thin-Layer Chromatography of Malondialdehyde

Whatman HP-K silica gel plates (10 × 10 cm; 250 μm thick) were washed by continuous development overnight in chloroform–methanol (1:1, v/v). Then 5-to 10-μL aliquots of the chloroform–methanol extract were streaked as a thin band, 5 mm long, along with 1-to 10-μL aliquots of malondialdehyde and TMP standards at concentrations ranging between 200 ng/mL and 100 μg/mL in chloroform-methanol (1 : 1), in separate lanes. The plates were dried at room temperature for 10 min, predeveloped in chloroform–methanol (1:1) to 10 mm from the lower edge of the plate, dried at room temperature, and then developed at 4 °C using hexane–acetone (80:20, v/v) as the mobile phase. Development proceeded until the mobile phase reached the top of the plate. After development, the plates were dried and sprayed with the TBA reagent and placed in a Frigidaire microwave oven at setting 9 for 10 min. The developed chromatograms were scanned in a Shimadzu CS-9000 spectrodensitometer in the transmission and fluorescence modes. For the former dual wavelength operation was used with the wavelength pair 532/570 nm. For the latter, 532 nm was used for excitation and a cutoff filter at 540 nm for emission. Scanning of the bands in the fluorescence mode was performed after allowing the plates to equilibrate at room temperature for 10 min. Loss of fluorescence was negligible for at least 30 min.

Gas chromatography–mass spectrometry

In another set of experiments, 100 μL of the chloroform–methanol extract was applied to Whatman HP-K plates as a thin band, 90 mm long. The plates were then dried, predeveloped in chloroform–methanol (1:1, v/v), and developed in hexane–acetone (80:20, v/v). The band comigrating with malondialdehyde standard was scrapped off the plate, extracted with chloroform–methanol–water (1:1:0.1, v/v/v), dried under N_2, and then reconstituted in 100 μL of hexane. Gas chromatography–mass spectrometry was performed with a Hewlett-Packard 5890 gas chromatograph interfaced with a Hewlett-Packard 5970 mass spectrometer.

RESULTS AND DISCUSSION

The thiobarbituric acid reactive components extracted from aerobically incubated rabbit testes homogenates are shown in the spectrodensitometer

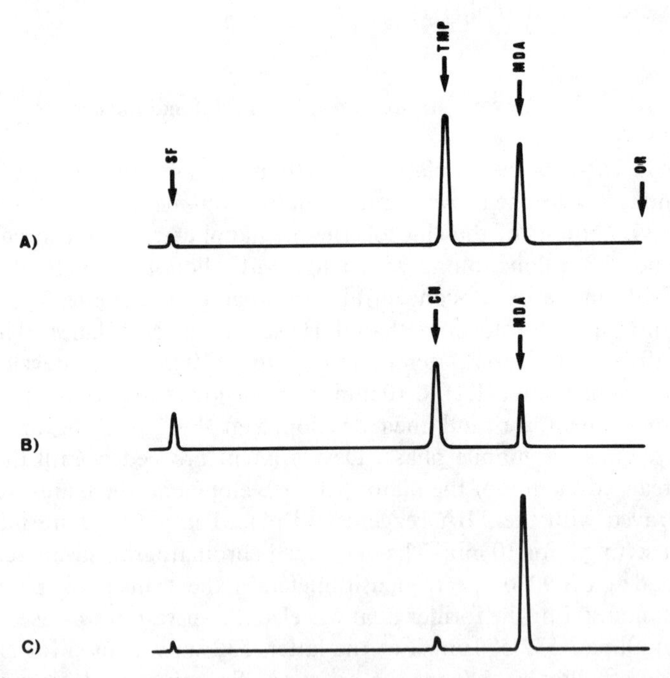

Figure 9.2. Spectrodensitometric traces of HPTLC plates with TMP and MDA standards (*A*) and the chloroform–methanol extract of aerobically incubated rabbit testes homogenates (*B* and *C*). Traces *A* and *B* were obtained in the transmittance mode. Trace *C* was obtained in the fluorescence mode.

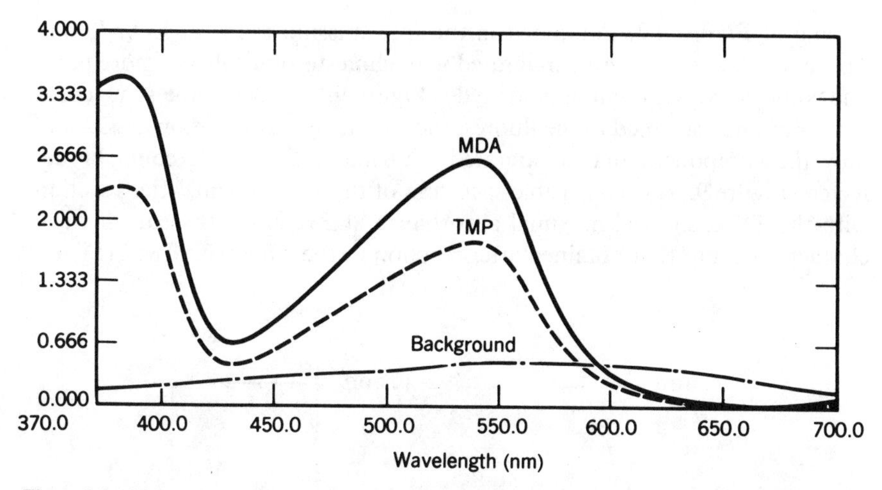

Figure 9.3. Absorption spectrum of the reaction product of aerobically incubated rabbit testes homogenates (MDA) with the TBA reagent. Also shown is the spectrum corresponding to TMP standard.

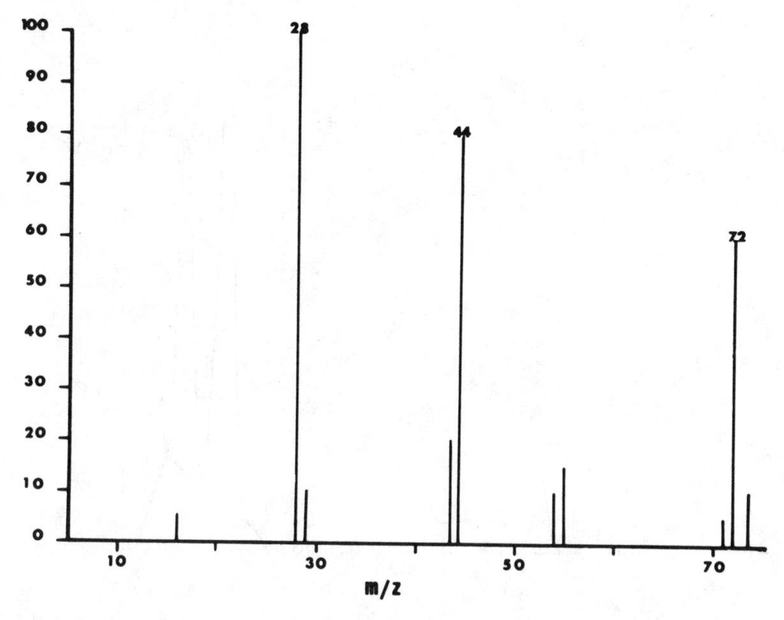

Figure 9.4. Mass spectrum of the malondialdehyde isolated from aerobically incubated rabbit testes homogenates.

tracing of Figure 9.2*B*, obtained in the transmission mode at 532/570 nm. The less polar component comigrated with cholesterol, while the more polar one comigrated with malondialdehyde (Figure 9.2*A*). When the same chromatogram was scanned in the fluorescence mode using excitation at 532 nm, only the component that comigrated with malondialdehyde could be detected (Figure 9.2*C*). The visible spectrum of this component, after reaction with the TBA, showed maximal absorbance at 532 nm with near identical characteristics to that obtained after reaction of the TBA with TMP (Figure

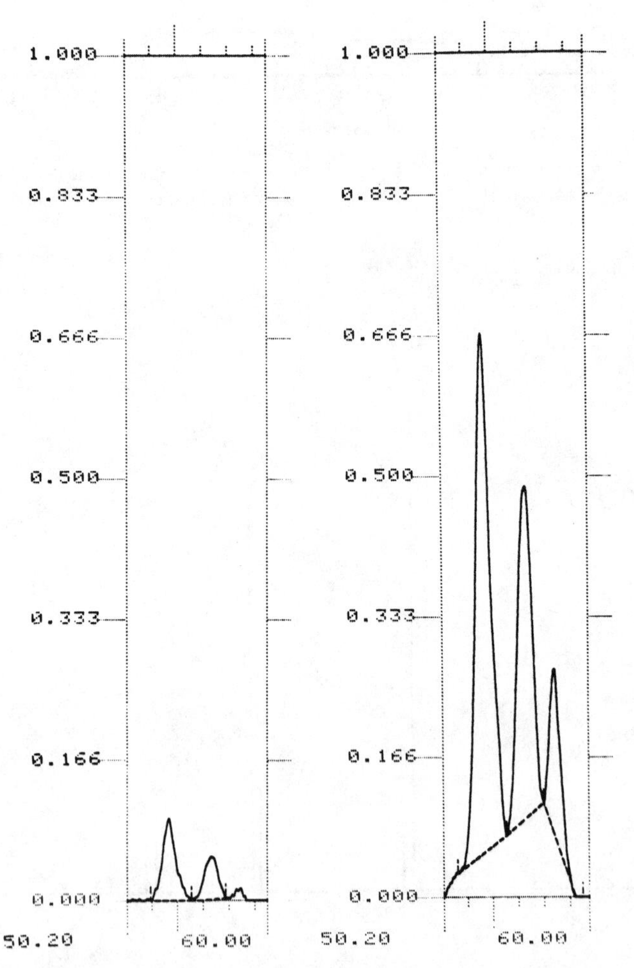

Figure 9.5. Spectrodensitometric traces of TMP hydrolysis products after spraying with the TBA reagent obtained in the transmitance (left) and fluorescence (right) modes.

9.3). Gas chromatography–mass spectrometry analysis of this component, after extraction by preparative HPTLC, indicated the presence of a single component yielding fragment ions m/z 28, m/z 44 (McLafferty rearrangement), and m/z 72 consistent with the structure of malondialdehyde (Figure 9.4). These results provide positive evidence for the identification of malondialdehyde as the only component reacting with the TBA in the band comigrating with malondialdehyde standard. Since scanning of the TBA–malondialdehyde product by fluorescence spectrodensitometry resulted in a 10-fold increase in sensitivity when compared with transmittance at 532 nm (Figure 9.5), and since the other component present in the chromatogram, that comigrated with cholesterol was barely detected in the fluorescence mode, we chose for the use of fluorescence spectrodensitometry as the method of choice for the determination of malondialdehyde. Separation time was 15 min for development to proceed to the top of the plate and the sensitivity of the system had a lower limit of 300 pg with linear detector response extended to 2 ng (Figure 9.6).

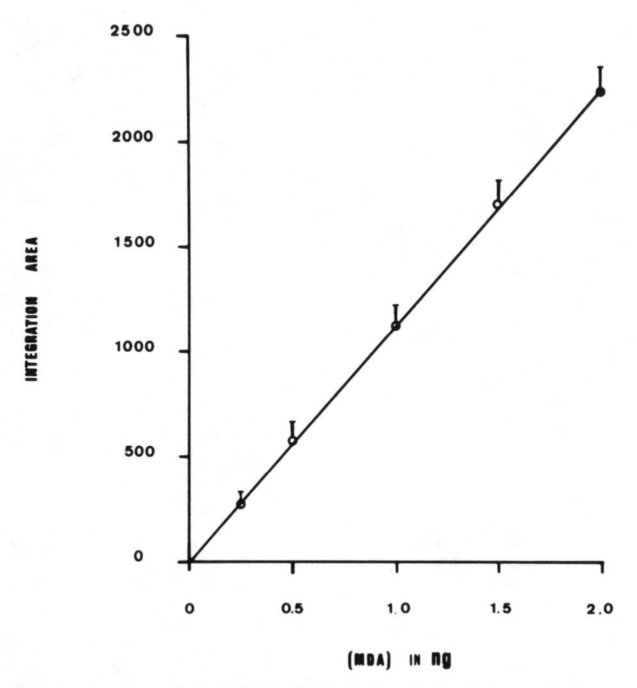

Figure 9.6. Standard curve of the Maldndialdehyde obtained from TMP acid hydrolysis generated in the fluorescence mode. The linear regression equation calculated through the origin had the form $y = 1095 \times$ ($r = 0.99$). Error bars are the standard deviations.

REFERENCES

1. L. K. Dahle, E. G. Hill, and R. T. Holman, *Arch. Biochem. Biophys* 98: 253 (1962).
2. H. I. Kohn and M. Liversedge, *J. Pharmacol. Exp. Ther.* 82: 292 (1944).
3. F. Bernheim, M. L. C. Bernheim, and K.M. Wilbur, *J. Biol. Chem.* 174: 247 (1947).
4. S. Patton and G.W. Kurtz, *J. Dairy Sci.* 34: 669 (1951).
5. R. O. Sinnhuber, T. C. Yu, and T. C. Yu, *Food Res.* 23: 626 (1958).
6. H. Ohkawa, N. Ohishi, K. Yagi, *Anal. Biochem.* 95: 351 (1979).
7. A. A. Barber, and F. Bernheim, *Adv. Gerontol. Res.* 2: 355 (1967).
8. J. M. C. Gutteridge, *FEBS Lett.* 128: 343 (1981).

CHAPTER

10

ANALYSIS OF ASCORBIC ACID BY THIN–LAYER CHROMATOGRAPHY*

JOSEPH C. TOUCHSTONE, TOM R. WATKINS, AND ERIC J. LEVIN

There is a need for ascorbic acid (AA) assays for biological samples. Current methods for this determination include titration with 2,6-dichloroindophenol (1), iodine and tetrachlorobenzoquinone (2), polarography (3), chemiluminescent detection (4), fluorescence detection after conversion to dehydroascorbate (5), and methods based on ascorbate oxidase (6). These methods are not ideal, having various disadvantages. More recently, a peroxidase-based colorimetric method was described (7). An HPLC method using detection at 220 nm has been proposed (8).

The need for a method was stimulated by early reports that an anti "nonspecific agglutination factor" was present in bovine seminal plasma which was stabilized by ascorbic acid (9). Human semen was also shown to have this factor (10). It has been reported also that total ascorbic acid levels were high in semen with a low percentage of motile spermatozoa (11). In contrast Videla (12) found no difference in ascorbic acid levels among oligo, azoospermic, and normal men. More recently, it was found that when ascorbic acid was added to bull sperm an increase in life span was seen (13). In another study, it was found that conception occurred in women whose partners, formerly infertile, took 1 g of ascorbic acid daily for 8 weeks (14). Subsequently, the following studies were undertaken to investigate the possible role of ascorbic acid in fertility problems.

During storage there is some conversion of AA to dehydro ascorbic acid. Thus, for determination of total ascorbic acid (TAA) content methods have been described for reduction to ascorbic acid (a) oxidation to the dehydro form (b) and use either as a measure of the total. In the method reported here copper salts are used to oxidize the AA to DHA and assay for TAA. In this way a more sensitive method results through the use of the ketonic functions

* Supported in part by grants from Hoffman La Roche Inc., Nutley, NJ.

for reaction with dinitrophenylhydrazine. The procedure described has a sensitivity of 5 ng of TAA.

The method described here utilized formation of the dinitrophenylhydrazone followed by separation by TLC and densitometry. High-performance liquid chromatography (HPLC) was also used. It was primarily developed for determination of total ascorbic acid in seminal fluid and blood. The modified method was necessary since previously mentioned reports gave no sample cleanup prior to chromatography. The cleanup increased specifivity since it enabled separation in the TLC. This method is reproducible, simple, and cost-effective . Since multiple samples can be separated on a single thin-layer chromatogram, the method has the potential of analyzing several hundred samples per day.

The dinitrophenylhydrazone of ascorbic acid was prepared by Ro and Kuether (15) and Lowry et al. (16) for determination of the vitamin in blood and urine. These were colorimetric methods that provided the basics for the present procedure in which the dinitrophenylhydrazone was isolated before quantitative evaluation by separation on TLC followed by densitometry. Separation before determination provided for more specific and reproducible results.

EXPERIMENTAL SECTION

Reagents

L-Ascorbic acid, trichloroacetic acid, thiourea, copper chloride, and solvents were all of analytical grade as obtained from Baker Chemical Co. 2,4-Dinitrophenylhydrazine (33% water) was obtained from Fluka Chemical Co. Radiolabled ascorbic acid ([L-^{14}C]carboxyl) was purchased from Amersham. This labeled compound was essentially pure, as it gave one peak on scanning with a Bioscan BID system 100 chromatogram scanner following separation by TLC as described in the methodology.

The reagents as prepared each day were 1% cupric chloride, 5% thiourea, 4% trichloracetic acid, and dinitropenylhydrazine (2% in 9 N sulfuric acid). For the derivatization 0.5 mL of 4% TCA was added to 0.5 mL of seminal fluid or 0.5 mL of plasma. Then 1 mL DPN reagent (20 mL DPNH + 1 mL $CuCl_2$ + 1 mL thiourea) was added. The mixture was heated 30 min in a boiling water bath, followed by cooling in an ice water bath. To the reaction mixture was added 5 mL of ethylacetate. Extraction was carried out by vortexing followed by centrifugation. From each reaction mixture a 5-μL aliquot of the ethylacetate (or 2 μL when 2 mL of ethyl acetate was used) was applied to the TLC.

The aliquot was applied to starting line of the TLC (20 × 20 cm) (Whatman LK5F) scored in 10-mm lanes. Predevelopment in a mobile of ethylacetate–chloroform (1:1) to the juncture of the preabsorbent with the sorbent facilitated alignment at the origin. After drying, the chromatogram was developed to 2 cm from the top of the layer, again with chloroform–ethylacetate (1:1). This was followed by densitometry as described below.

Instrumentation and Chromatographic Evaluation

TLC

A Kontes Model 800 fiberoptic scanner was used. The light source was equipped with a filter having maximal output at 440 nm. The instrument was operated in the double-beam mode. Output was integrated with a Hewlett-Packard model 3390A instrument. A standard curve was prepared by adding serial amounts of ascorbic acid to individual 15-mL test tubes and the procedure–derivatization, TLC, and densitometry was carried out. The regression line found for this procedure is shown in Figure 10.1.

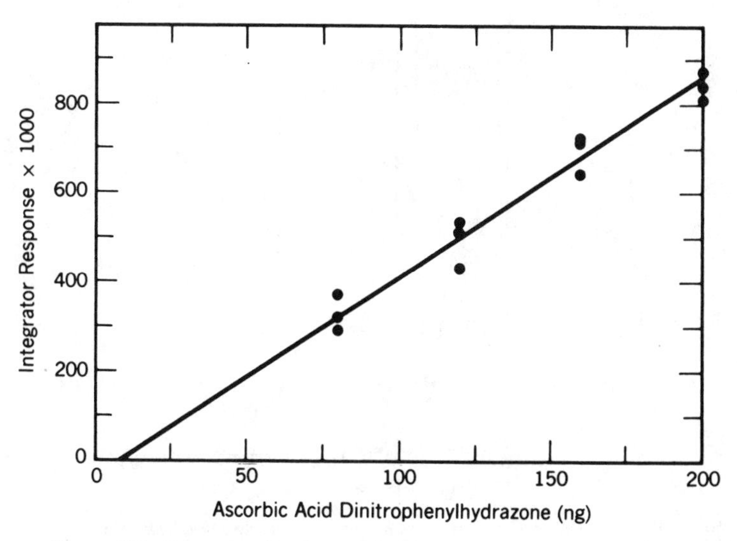

Figure 10.1. Regression line for serial amounts of ascorbic acid DNPH.

HPLC

The HPLC unit was a modular apparatus consisting of an LDC pump, a Kratos spectroflow 773 variable wavelength detector, a Reodyne loop injector, with a Whatman Partisil 10 (ODS3) column coupled with a Co-Pell

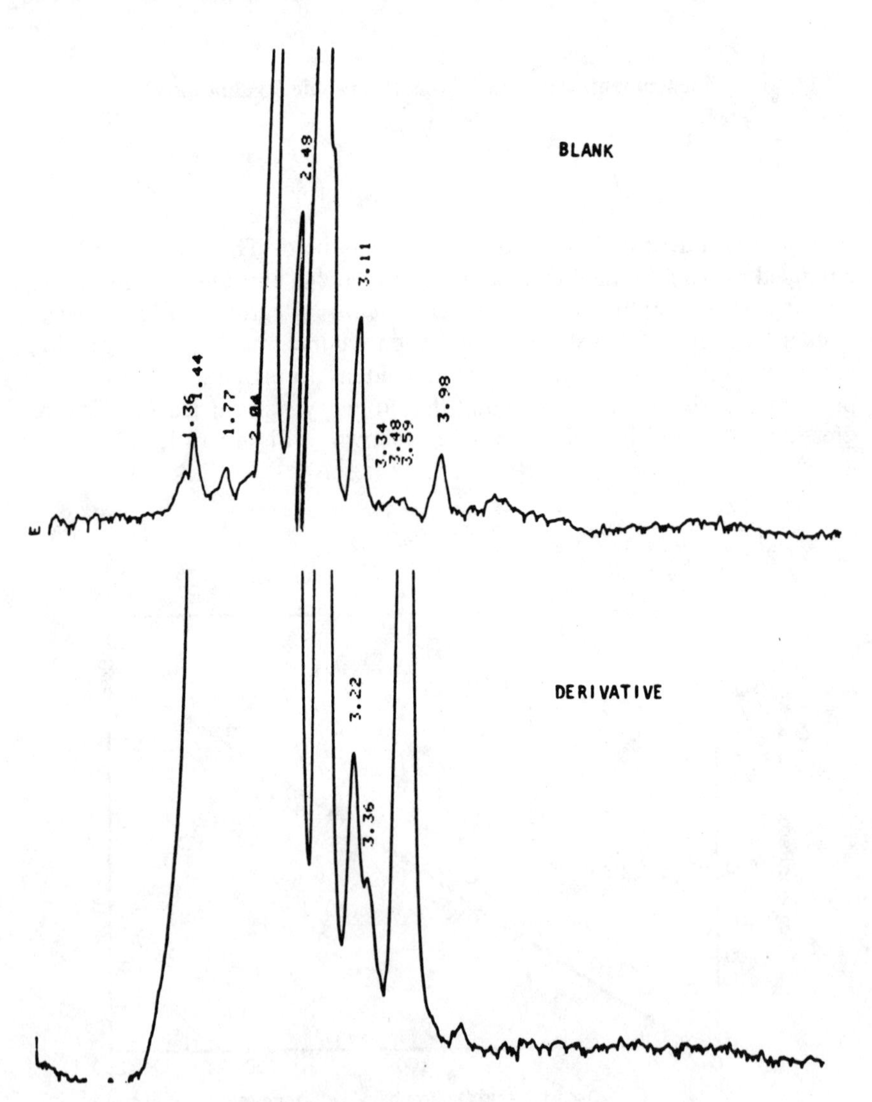

Figure 10.2. High-performance liquid chromatogram showing separation of ascorbic acid dinitrophenylhydrazone. See text for conditions.

guard column. The mobile phase was acetonitrile–water (80:20) with a flow rate of 1 mL/min. A wavelength of 254 nm was used for scanning.

RESULTS AND DISCUSSION

Figure 10.2 shows the high-performance liquid chromatogram obtained for the dinitrophenylhydrazone of ascorbic acid prepared by the method de-

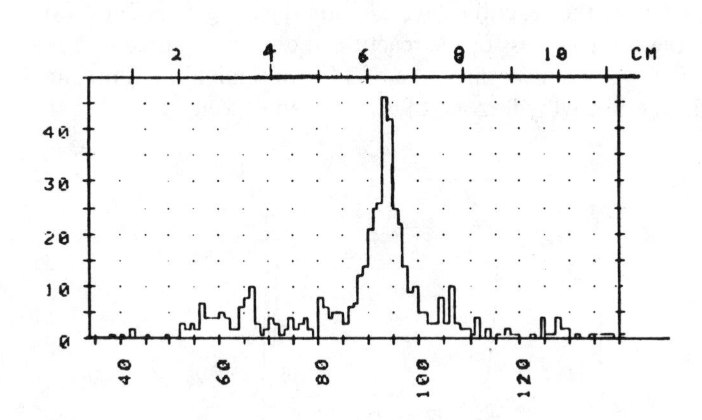

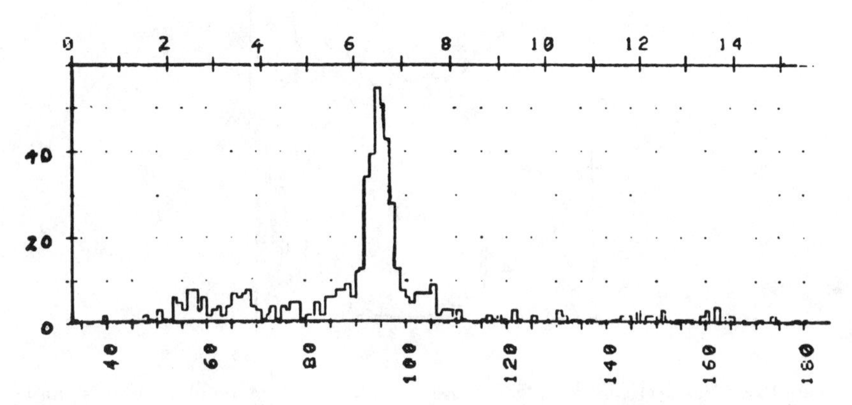

Figure 10.3. Recovery of ascorbic acid (^{14}C) from plasma after the described procedures.

scribed. This was separated using a Whatman Partisil 10 (ODS-3) reversed-phase column (25 × 4.6 mm) with a CoPell guard column. The results show a ready separation of the ascorbic acid dinitrophenyl hydrazone. It was not possible to continue using this method since the acidic reaction mixture was deleterious to the column. Consequently, the TLC method was preferred. A procedure to prepare the derivative for routine use in HPLC is under investigation. More recent developments in sorbent technology indicate that newer columns may be amenable to this use being less effected by acid.

Figure 10.3 shows the radioscan of a TLC of radiolabeled ascorbic acid added to plasma and a standard and carried through the procedure. As indicated in the figure, the reaction gave essentially a single product. Both results indicate the completeness of the reaction and the ease of scanning for quantitation. Figure 10.4 shows the scans of TLC of seminal fluid prepared by the method. The phenylhydrazone of AA is readily seen.

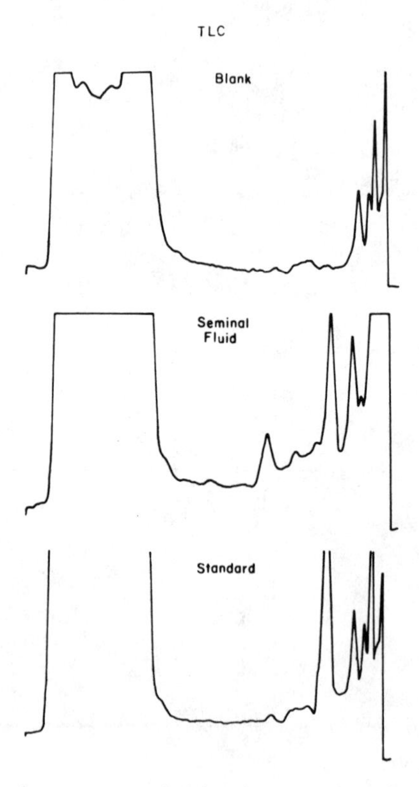

Figure 10.4. Thin-layer chromatograms showing the separation of ascrobic (as DPNH) from seminal fluid.

Table 10.1. Effect of Ascorbic Acid Ingestion on Blood and Semen Levels of Ascorbic Acid

	Plasma and Semen Levels[b] (mg/mL)							
	Control Day		1 Week		2 Weeks		3 Weeks	
Patient[a]	P	S	P	S	P	S	P	S
1	2.3	42	2.5	64	6.5	48	7.5	38
2	1.5	58	3.5	50	ND	84	5.5	44
3	ND	78	2.0	84	2.5	85	3.0	80
4	ND	52	3.0	74	2.5	50	ND	ND
5	ND	60	5.0	120	5.0	80	80	174
6	3.2	56	4.0	66	4.5	80	8.0	200

[a]Each patient ingested 400 mg/day of ascorbic acid during the experiments (100 mg/four times/day).

[b]P, plasma; S, semen; ND, not done.

Recovery of radiolabeled ascorbic acid added to blood plasma and seminal fluid was 92% after derivitization, extraction, and thin-layer chromatography followed by scanning using a Bioscan isotope scanner.

Four replicate determinations of ascorbic acid in aliquots of the same sample of seminal fluid gave a mean of 145 ng/mL with SD of 6.1 and CV of 4.2.

Table 10.1 gives the results obtained for seminal fluid and blood obtained from volunteers after ingestion of different amounts of vitamin C tablets. The levels of the vitamin did not improve appreciably as result of the intake.

Although the dinitrophenylhydrazone of ascorbic acid is readily prepared, its separation on LC is not practical since the acid conditions resulting from the reaction is deleterous to the column. To use LC it would be necessary to perform further cleanup before the analysis. The sensitivity of the detector used in our LC system (Schoeffel SF770) does not appear to match that of the TLC methodology. The TLC method has a level of detection of 5 ng per applied sample provided the derivative is well separated.

At this stage, it appears that the TLC method is the method of choice.

REFERENCES

1. D. W. Hughes, *J. Pharm. Sci.* 72: 126 (1983).
2. N. K. Pandy, *Anal. Chem.* 54: 793 (1982).

3. P. W. Alexander and H. Marphaung, *Talanta* 29: 213 (1982).
4. J. E. Frew and P. Jones, *Anal. Lett.* 18: 1579 (1985).
5. R. B. Roy, A. Conetta, and J. J. Salpetir, *Assoc. Off. Anal. Chem.* 59: 1244 (1976).
6. C. W. Bradberry and R. N. Adams, *Anal. Chem.* 55: 2439 (1983).
7. R. Q. Thompson, *Anal. Chem.* 59: 1119 (1987).
8. L. L. Lloyd, McConvile, F. P. Watner, J. F. Kennedy and C. A. White, *LC GC Magazine* 5: 338 (1987).
9. P. E. Lindahl and J. E. Kihlstrom, *Fertil. Steril.* 5: 241 (1984).
10. N. J. Bullimore, J. P. Crich, and A. M. Jequire, *Andrologia* 13: 387 (1981).
11. P. Koets and L. Michelson, *Fertil. Steril.* 1: 15 (1956).
12. E. Videla, A. M. Blanco, M. E. Galli, and E. Fernandez-Collazo, *Andrologia* 13: 212 (1980).
13. V. Waganathan, *Ind. Vet. J.* 47: 1059 (1970).
14. W. A. Harris, T. E. Harden, and E. R. Dawson, *Fertil. Steril.* 32: 455 (1979).
15. J. H. Roe C. A. Kuether, *J. Biol. Chem.* 147: 399 (1943).
16. O. H. Lowry, J. A. Lopez, and O.A. Bessesy, *J. Biol. Chem.* 160: 609 (1945).

CHAPTER

11

ONE- AND TWO-DIMENSIONAL SCANNING FOR ^{32}P AND OTHER UNCOMMON TAGS

EDWARD RAPKIN

Scanning thin-layer chromatograms (TLC) and paper chromatograms for content of radioactivity is a topic once of active (1960–1970) and then of diminished interest (1970–1980), when it became evident that instrumentation could not keep up with measurement requirements. But, since 1980, there has been revived interest as new instrument concepts make possible increased throughput, greater accuracy, and 2-D imaging. Nowhere is the pressure for something better as pronounced as it is for the measurement of ^{32}P and other energetic beta-emitting isotopes which are not easily measured by established methods. Ever-present in any consideration of instrumentation for radioactivity scanning is the competition of autoradiography; to be of interest, a scanning technique must offer advantages, not only technical, but also economic.

1-D SCANNING WITH GEIGER AND PROPORTIONAL COUNTERS

The first practical method of scanning planar 1-D chromatograms for beta activity was with windowless-flow Geiger counters having narrow slit collimators. Most often the sample was passed under a stationary detector, but in some instances the sample was fixed and the detector moved. With the position of the sample relative to the detector known, and usually with the measured count rate driving a strip-chart recorder via a ratemeter, it was possible to establish quite accurately the location of activity on the chromatogram. Such devices measured ^{3}H with 1–2% efficiency and ^{14}C with perhaps an efficiency of 15% or more; with the sample in close proximity to the slit window, resolution was essentially a direct function of the width of the slit.

Later, for no really obvious reason, windowless proportional counters replaced Geiger counters; mechanical design remained essentially unchanged. With proportional counters, backgrounds were lower than with

Geiger counters, but so was counting efficiency. It is true that proportional counters could count at higher rates than could Geiger counters, and that proportional counters offered the prospect of pulse-height analysis with the almost always unrealized potential for isotope discrimination, but proportional counters added complexity and cost to the electronics and were rarely needed for the high activity levels where their use is mandatory. As with Geiger counters, resolution was a function of collimator slit width.

Though these early mechanical Geiger and proportional counting systems provided digital information to complement their accurate positional information, they suffered from a major shortcoming; they could examine only a small segment of the chromatogram at any one time. Throughput could be achieved with fast scans, but fast scans did not allow low-level measurement. Autoradiography with samples run in parallel became the method of choice for many, even though exposure times were substantially longer than was the time needed for a mechanical scan. While autoradiography, inherently a 2-D process, does not easily provide quantitative information, spot locations can be known accurately, plates can be scraped or paper cut into pieces, and quantitative results can be had via liquid scintillation counting.

1-D SCANNING WITH POSITION-SENSITIVE PROPORTIONAL COUNTERS

Adaptation of the position-sensitive proportional counter (PSPC) of Borkowski and Kopp (1) and of Charpak (2) to the counting of thin-layer plates and paper chromatograms, first suggested by Borkowski and Kopp (3,4) and later by Gabriel and Bram (5), and implemented by the Numelec (France), Berthold (Germany), and Bioscan (U.S.) companies, provided an interesting solution to the principal problem—lack of speed—of the early mechanical scanners. Most PSPCs have an open-sided detector chamber perhaps 20 cm long × 1 cm wide × 0.5 cm high; the chamber walls serve as the cathode of a proportional counter. The single anode wire—20 cm is the normal length—is positioned parallel to and in close proximity with a 1-D track on a chromatogram; it looks at the entire track at one time (Figure 11.1). P-10 counting gas (90% argon/10% methane) is most often used.

The PSPC does have limitations, but to understand them the detection process must be understood. At a radioactive site on the chromatogram, beta particles are emitted isotropically, that is, in a uniformly spherical distribution. Unless there is some backscatter, at most only half the particles, those directed toward the counting gas above the sample, might be measured; the others give up their energy to the sample or its support. But, in fact, the situation is far less happy. Many particles emitted in the direction of the

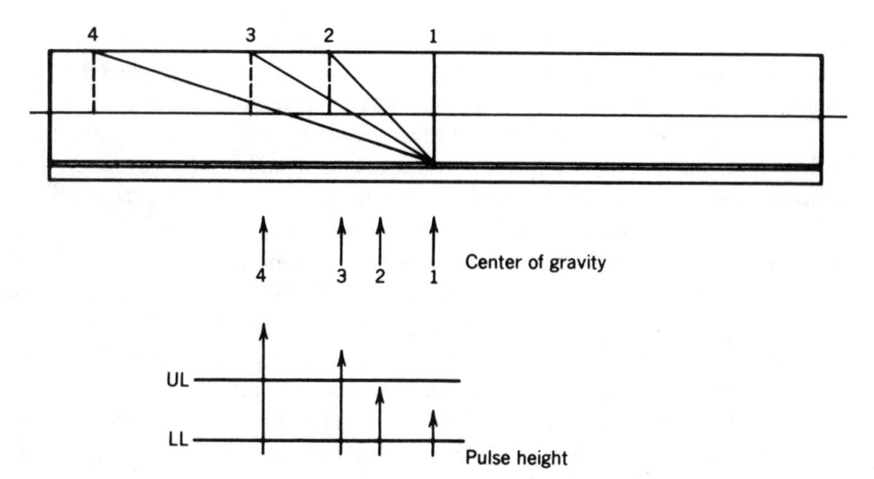

Figure 11.1. Diagram showing how the wire looks at the entire track in the layer.

chamber never make it out of the sample (self-absorption), many do not have sufficient energy to ionize gas, and many give pulses that must be rejected because their origin is ill defined. This last requires further discussion.

A beta particle emitted into the counting gas moves in a straight line, its path essentially uninfluenced by electric fields within the counting chamber since the typical operating voltage (1200–1800 V) is low relative to the energy of the particle. As the primary particle traverses the gas and interacts with it, gas ionization results and the particle gives up some or all its energy. If all of the energy is not lost to the counting gas, the particle eventually strikes the conducting wall (cathode) of the counting chamber where, unless there is backscatter, all residual energy is surrendered. As the primary particle has traveled, it has left a linear track of secondary electrons and ions.

Secondary electrons drift to the positively charged anode wire in a plane perpendicular to the wire, while ions more slowly drift to the cathode. The anode has a narrow cross section, 25–50 μm, and the surrounding space-charge density is high. As the electrons near the anode they are increasingly accelerated by the internal electrical fields and they interact with additional gas molecules, resulting in further ionization (gas multiplication), and ultimately all secondary electrons are incident upon the anode where they deposit their collective charge.

In commercial PSPCs, two different systems have come into use to determine the position of the charge upon the wire. One approach employs a high-resistance anode wire, made by flame deposition of a carbon film on a quartz fiber. For each event the charge generated at either end of the wire is examined; with the wire having uniform conductivity, the ratio of the amplitudes of the two pulses indicates the center of gravity of the charge. The alternative method employs a conducting anode, often a gold-plated tungsten wire, which is inductively coupled to a delay line, which, itself, may make up a wall of the chamber. A pulse on the anode wire results in pulses at either end of the delay line; measurement of the time differential between the delayed pulses provides positional information.

But, there are problems. While the maximum-energy ^{3}H particle (18 keV) penetrates only 0.3 mm of counting gas at atmospheric pressure before all energy is given up, the range for ^{14}C ($E_{max}=156$ keV) is many centimeters and that of ^{32}P ($E_{max}=1710$ keV) is greater than 10. What this means is illustrated in Figure 11.1; no matter what the angle of incidence of the ^{3}H beta, the apparent location is a reasonably accurate representation of fact. However, the same is less true for ^{14}C and still less for ^{32}P. Energetic events with low angles of incidence create long tracks with substantial ionization whose center of gravity may be far removed from the site of the decay event. It should be evident that the more energetic the event or the further away from the perpendicular, the less accurate will be the location.

There are at least partial solutions to this dilemma. The mechanical solution is to use stacked multiple-slit collimators to stop angular components. But collimators decrease counting efficiency, thereby negating some of the advantage of examining an entire track at one time. Also, they may cast a "shadow" on the record, sometimes making one point seem as two, especially if the collimators are relatively thick as they must be to stop ^{32}P. More usual than the collimator is a rather unique system of pulse-height analysis.

Of the emissions from a sample site, the ones that can be most accurately determined are those that enter the chamber and traverse a plane perpendicular to, or near perpendicular to, the sample at the activity site and at right angles to the anode wire; they travel the shortest path, create the shortest track, and give the smallest pulses. The centers of gravity of such events fall close to their true locations. As the angle of entry decreases the tracks become longer, there is more gas ionization, and pulses are larger; the center of gravity of the secondaries incident upon the anode wire falls farther away from the origin (Figure 11.1).

Pulse-height analysis can usefully eliminate unwanted counts. Small events are recorded, large ones are rejected. With a high-resistance anode wire, pulse amplitude is obtained from the cathode; for delay-line systems the pulse on the anode wire is directly measured. But, as with collimators, pulse-height analysis reduces counting efficiency, sometimes substantially. While nearly half of the ^{32}P events from a spot on a TLC plate might be detected, over 90% rejection is required to obtain any useful resolution with a PSPC. Even then, backscatter can lead to further inaccuracies.

The counting chamber of a PSPC is so small, and most ^{32}P events are so energetic, that there is considerable backscatter. If the backscatter reenters the counting chamber at a low incident angle, it will likely create an another long track whose record will be rejected by pulse-height analysis. But, if the scatter takes a short path through the chamber a small pulse will result and will be recorded; needless to say, its record will be in error since its origin is happenstance. The net effect of this, especially for ^{32}P is a diffuse data presentation. Not only is resolution poor, but there may be an overall background which cannot be explained by activity on the sample.

Still another problem of the PSPC is seen only at high count rates. While a proportional counter can handle high rates, when an entire track is examined at one time with what is in effect but one counter, it is quite possible to exceed the counting capacity of that counter. Should two events occur at different locations along the track but within the resolving time of the counter they will be seen as a single event, and in the wrong location. If the pulse they generate is sufficiently large, perhaps they will be rejected. But, if small, or if ^{3}H is being counted and pulse height analysis does not come into play, the record of that pulse, and others arising from pileup in the same way, will be in error.

It should be a simple matter to test the performance of a PSPC for TLC scanning. But, until now, standard plates have not been available and the tendency has been to compare plate scanners with autoradiographs, which, themselves, do not give digital information; a properly adjusted PSPC shows ^{3}H spot location about as well as an autoradiograph, does a little less well with ^{14}C, and probably gives no useful results for ^{32}P, but it is almost always much simpler and faster. A more objective test would be helpful.

Amersham, the radioactivity sales arm of the U.K. atomic energy organization has developed a test plate for measuring the performance of TLC radioactivity scanners. It is capable of checking response uniformity, the integrity of positional information, resolution, and response to different activity levels. It is constructed by cementing plastic strips containing ^{14}C activity into depressions milled at selected locations on a 20×20-cm aluminum plate (Figure 11.2); the activity is uniform across each strip, but in the zone centered on 100 mm some strips have different levels of activity than others. The plate may be treated as if it were made up of five 1-D tracks, with each track measured separately; results obtained with a PSPC are shown in Figures 11.3*a–e*. Some comments are in order.

Figure 11.3a: Track 1 (20 mm). Two activity strips at 30 and 170 mm are present. The PSPC finds their location quite accurately, though it suggests a strip width at baseline of at least 10 mm whereas each strip is actually about 1 mm wide.

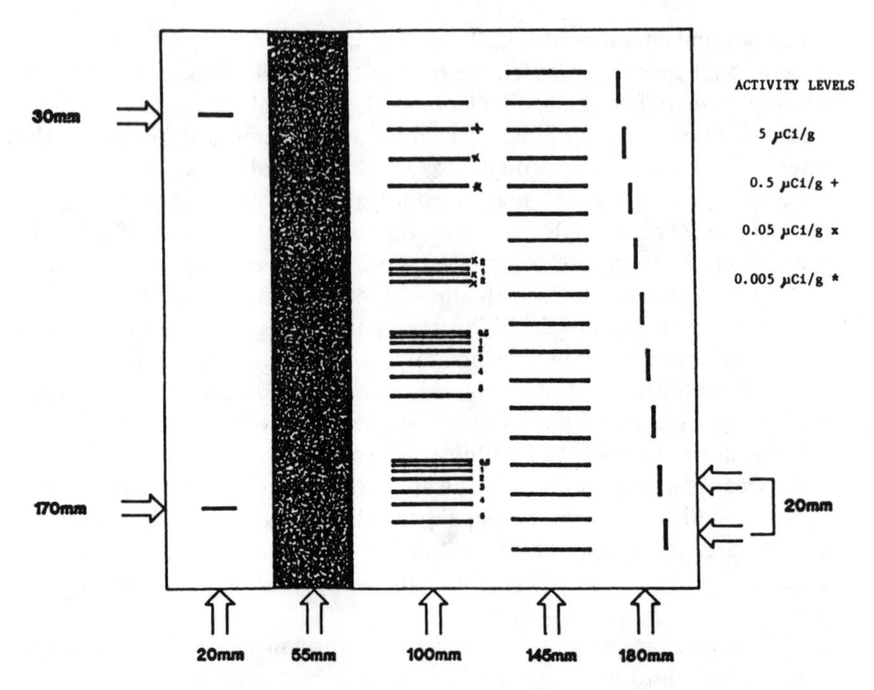

Figure 11.2. Diagram showing configuration of a ^{14}C calibration plate for linear analyzers.

Figure 11.3b: Track 2 (55 mm). A uniform activity band is a more difficult test for a PSPC. This one fails rather badly.

Figure 11.3c: Track 3 (100 mm). Another severe test of the PSPC combines measurement of resolution and response to different count rates. The two lowermost arrays each have seven strips but only six were found. The two closest strips are only 0.5 mm apart and such separation is beyond that claimed for such scanners and at this time cannot reasonably be expected. Instrument manufacturers do suggest that 1-mm resolution may be possible, but unless one knew for a fact that there were active zones 1 mm apart, it would only be a guess as to whether or not the regions between 41–45 and 88–92 mm be treated as one zone of activity or two. Each strip in both arrays has the same activity level; from the amplitude of the peaks it is clear that no useful quantitative information can be gotten from so closely spaced zones of activity.

With four strips spaced 1 and 2 mm apart, but with one internal strip and the extremities having only 1/100th of the activity of the central strip, the PSPC saw only a single peak but with the breadth of the four strips; possibly

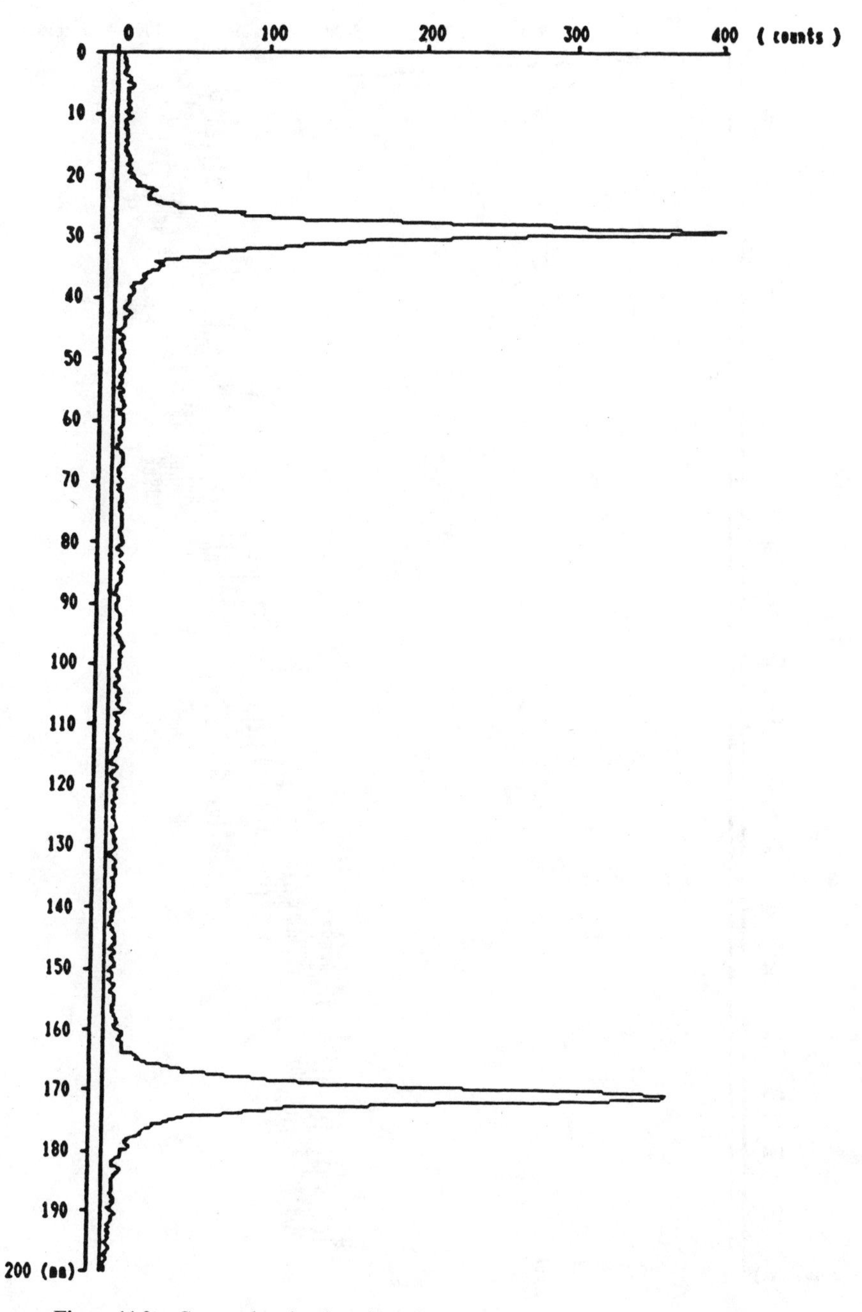

Figure 11.3a. Curves showing location of two zones of activity (at zonal 170 mm).

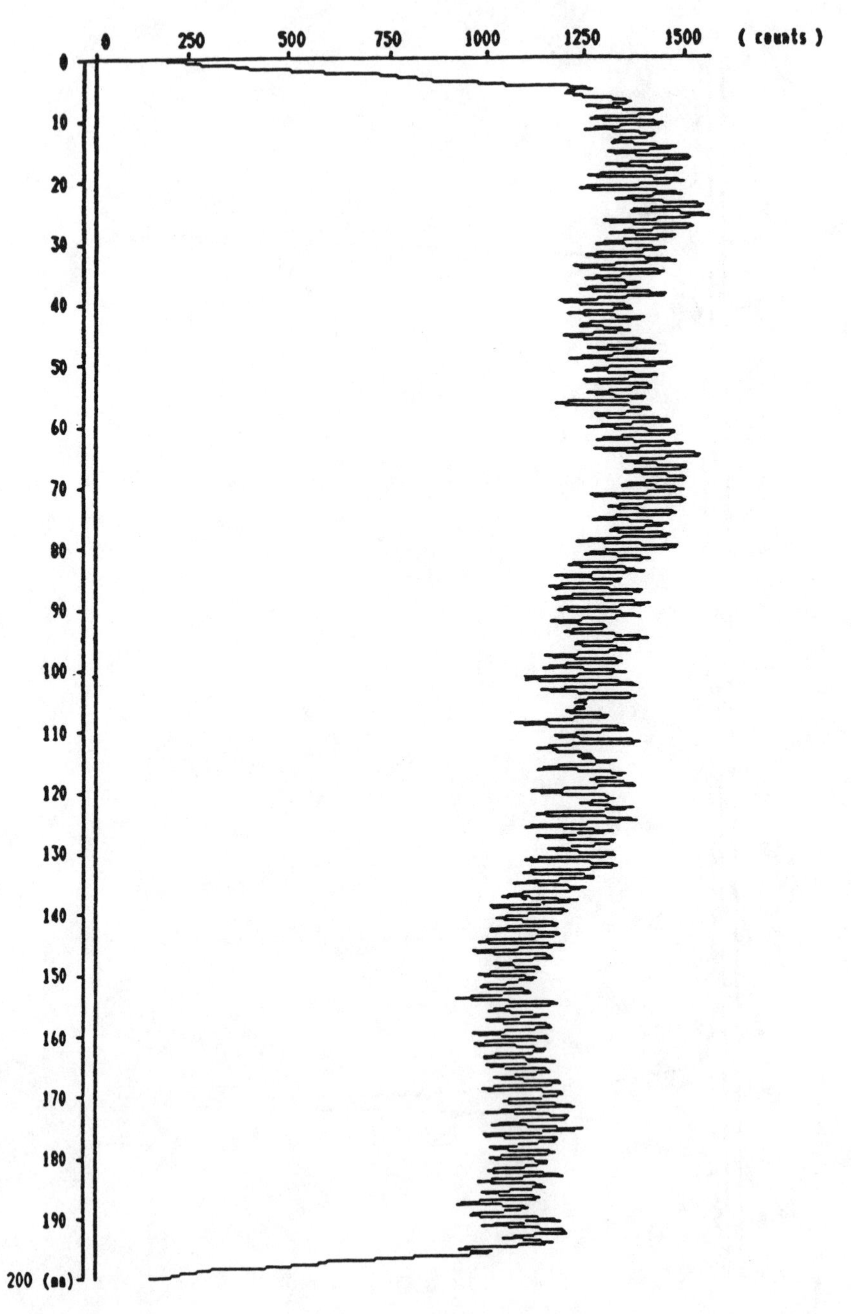

Figure 11.3b. Baseline scan.

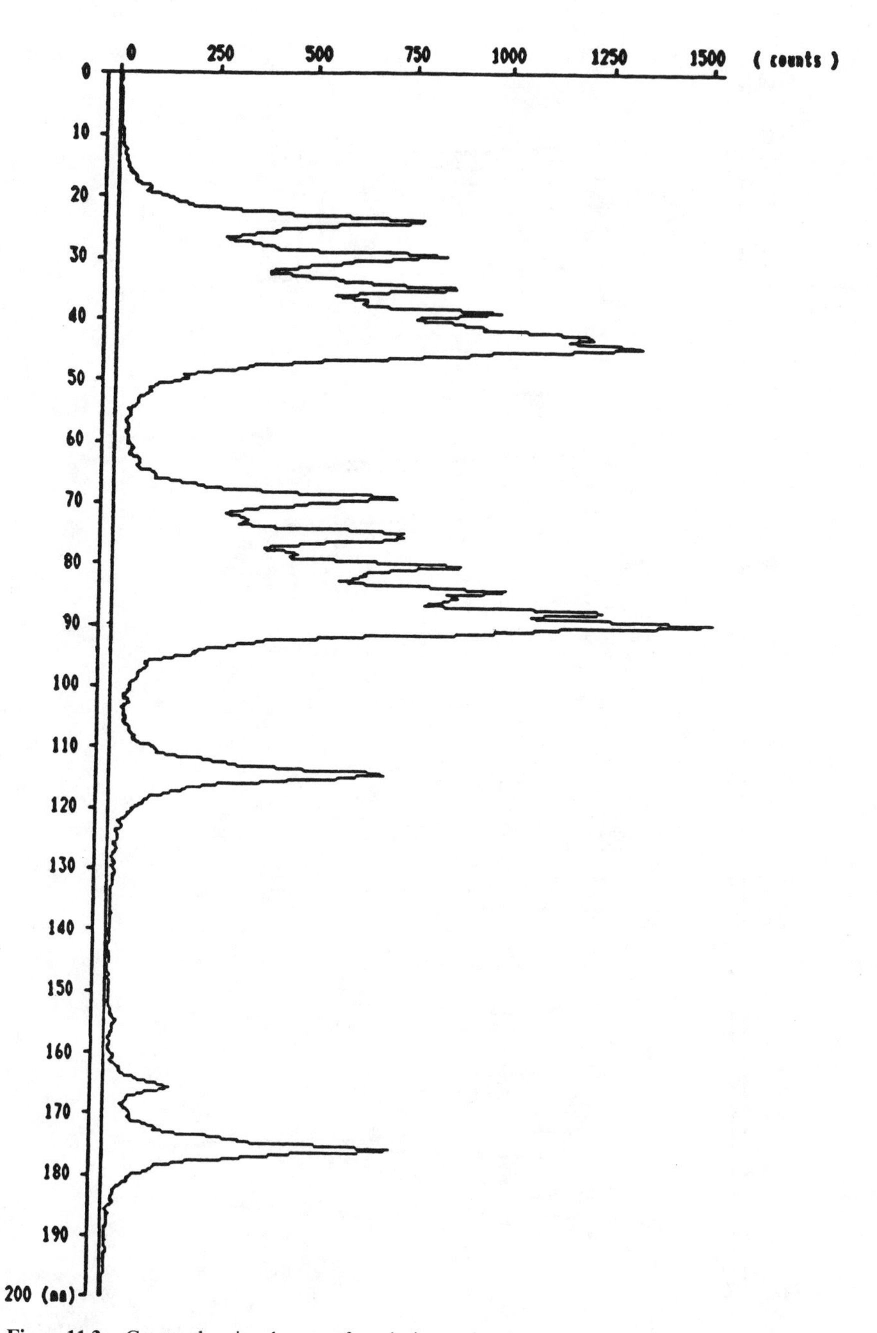

Figure 11.3c. Curves showing degrees of resolution and response to different count rates. See text.

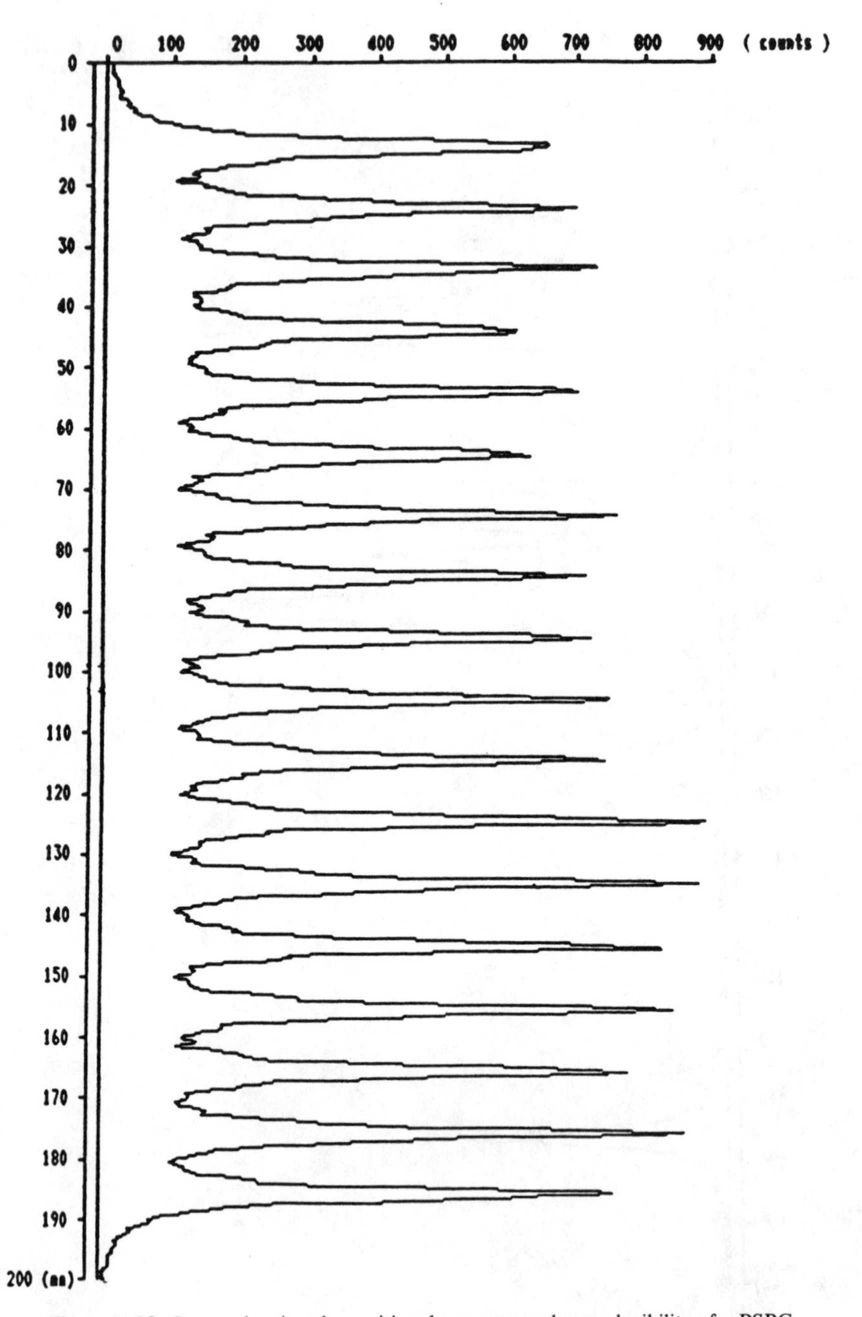

Figure 11.3d. Curves showing the positional accuracy and reproducibility of a PSPC.

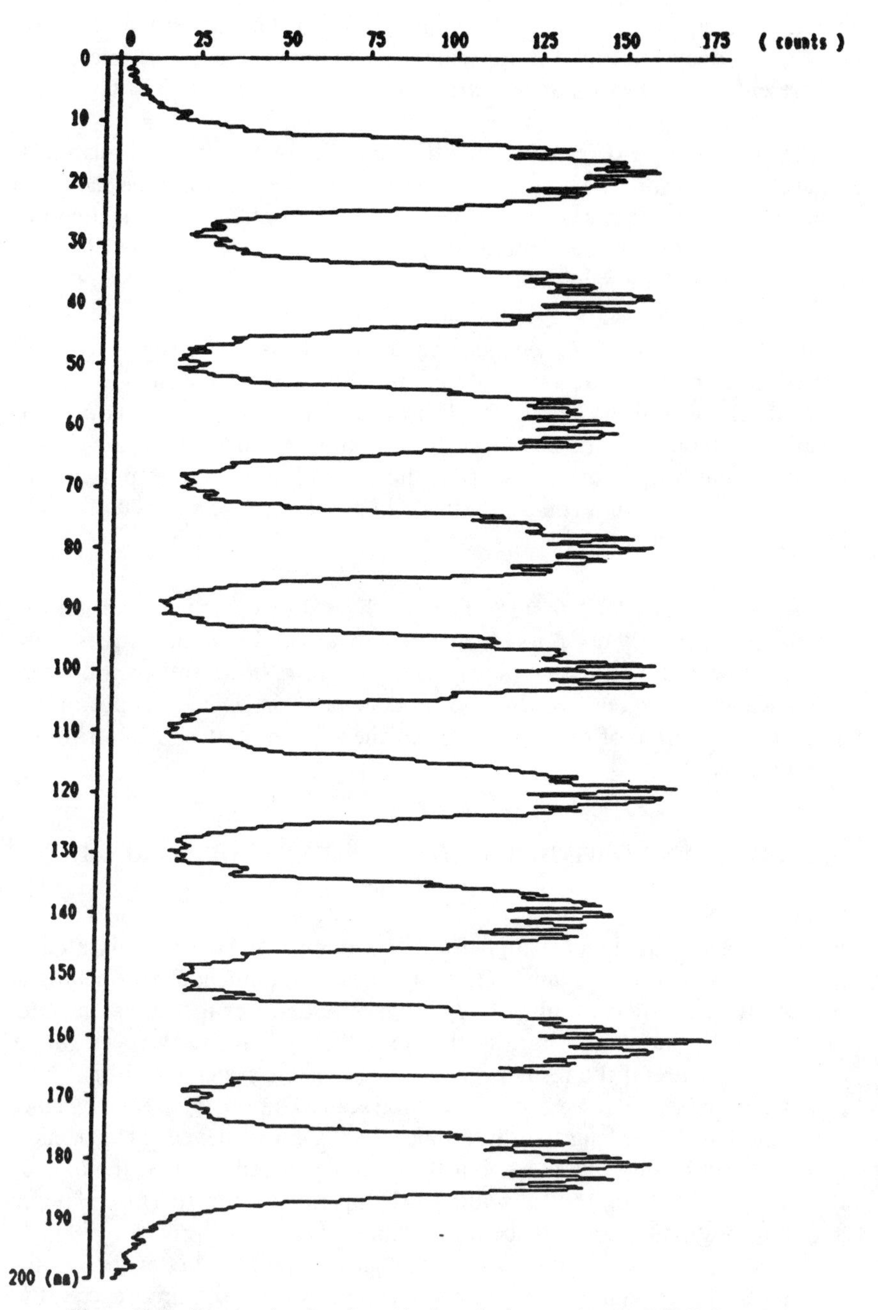

Figure 11.3e. Results showing the reviewing of a layer using the PSPC. See text.

counting time was inadequate. It should be noted that the apparent amplitude of the one peak found is approximately that which might be inferred to be present in the most isolated strip (5-mm resolution) of the two arrays previously discussed.

The uppermost array has four strips of different activity levels, each separated by 5 mm. Here, two peaks and perhaps a third can be seen; the fourth, and lowest activity, is completely lost, but possibly only as a consequence of the form of data presentation which cannot be expected to cope with so wide a dynamic range.

Figure 11.3d: Track 4 (145 mm). Eighteen evenly spaced strips on 10-mm centers test positional accuracy along an entire track. Here, it is seen that such accuracy is quite good; the PSPC can obviously be used for spot location if spots are not too close to one another and activity levels are relatively uniform. In other tests (not shown) with more severe pulse rejection, baseline resolution can be attained but at the expense of less uniformity between peak contents.

Figure 11.3e: Track 5 (180 mm). The nine bars should each show a flat top but this is apparently not possible due to statistics and nonuniform response over narrow sections of the anode wire. This test shows rather good performance when it is realized that the anode wire of the detector is positioned along the centerline of the track but that the first and last pair of strips are considerably offset.

2-D SCANNING WITH POSITION-SENSITIVE PROPORTIONAL COUNTERS

PSPCs are being used increasingly for 2-D scanning. 2-D scans are treated as many closely spaced 1-D lanes. Often, the usual 10-mm chamber opening is reduced with an aperture plate to 3–5 mm. The detector measures one lane and then it or the sample steps to the next. After all lanes have been run, a composite picture of the results is presented; in some systems with real-time display, it is possible to see a top view of the scan as it progresses lane by lane. Resolution along the length of the anode wire is a consequence of electronics, including pulse-height analysis, but resolution perpendicular to the axis of the wire is at best equal to the width of the opening of the aperture plate; X and Y resolutions are almost certainly different.

The narrower the opening and the smaller the step, the better the resolution but the fewer counts recorded per unit time. Steps must never exceed the width of the aperture or sections of the plate will not be scanned, They may be less than the width of the opening and this will increase the number of counts observed. Peak positions are accurately determined but unless severe pulse-height analysis is practised to the substantial detriment of counting efficiency, zones of activity appear much larger than they really are because the detector sees and records what might be described as an inverted cone at each active point; point sources as far apart as the chamber opening appear as two peaks from within a single envelope of activity, with the saddle between them having substantial amplitude. Here, with the possibility of several peaks growing out of a single base, an isometric projection or a color printout of the topographic view can facilitate visualization and interpretation of results.

VANGUARD 1-D/2-D SCANNER

With there being an increasing requirement for scanning ^{32}P and other energetic isotopes, the Vanguard scanner (Figures 11.4*a,b*) has been developed in recognition of inherent problems in counting such isotopes with the PSPC. The Vanguard represents a return to fundamentals; it is a modern-day Geiger counter with mechanical collimation. To increase speed it incorporates a detector array with 10 miniature, windowless Geiger detectors in a single row on 20-mm centers (Figure 11.5); windowless operation is requisite for ^{3}H measurement, though much less a requirement for more energetic isotopes. The array moves stepwise over the sample—which may be a TLC plate, one or more paper strips, wet or dry gels, or a blot—scanning either 10 lanes at a time or one lane in 10 locations in either *X* or *Y* directions and then, once having reached the end of its pass, translating *Y* or *X* and scanning through the next lane in the reverse direction.

Figure 11.4a

Figure 11.4b

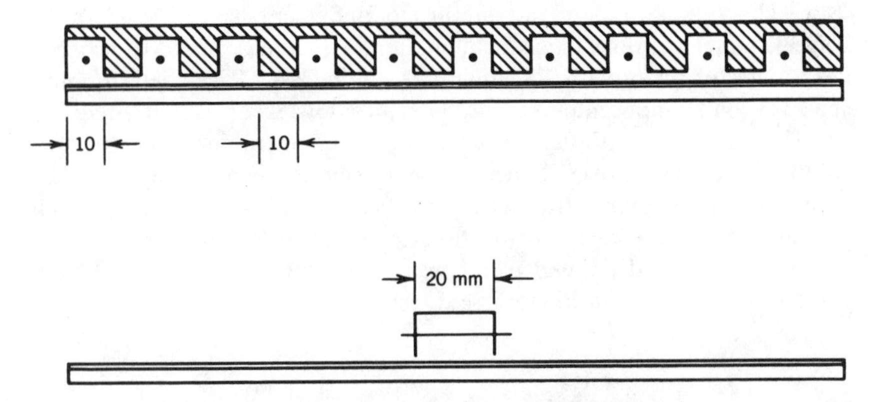

Figure 11.5. Diagram showing detector arrays with ten minature windowless Geiger detectors.

Steel collimators of various sizes and thicknesses (Figure 11.6) are available and are held in place magnetically at the underside of the array; the distance of scan steps and X and Y translation is in accord with the dimensions of the collimator openings. Resolution is essentially a function of the size of the collimator openings, though it is also affected by the distance between the sample and the collimator (Figure 11.7); in recognition of this fact, the Vanguard scanner includes provision to raise and lower the sample table to compensate for different collimator and sample thicknesses. Inasmuch as the number of counts recorded is an inverse function of collimator size, a balance must be struck between resolution and sensitivity.

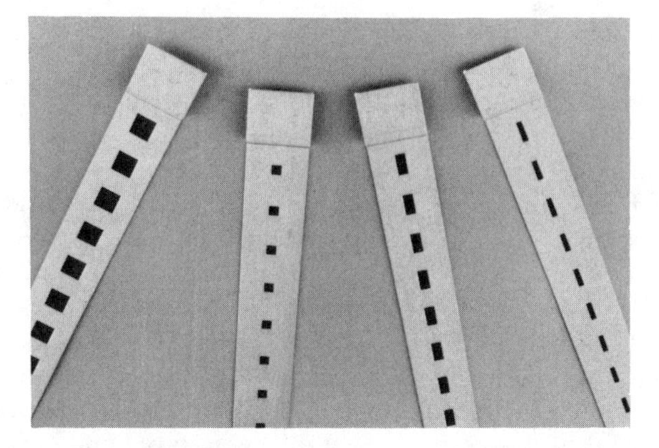

Figure 11.6. Steel collimators used to control resolution. See text.

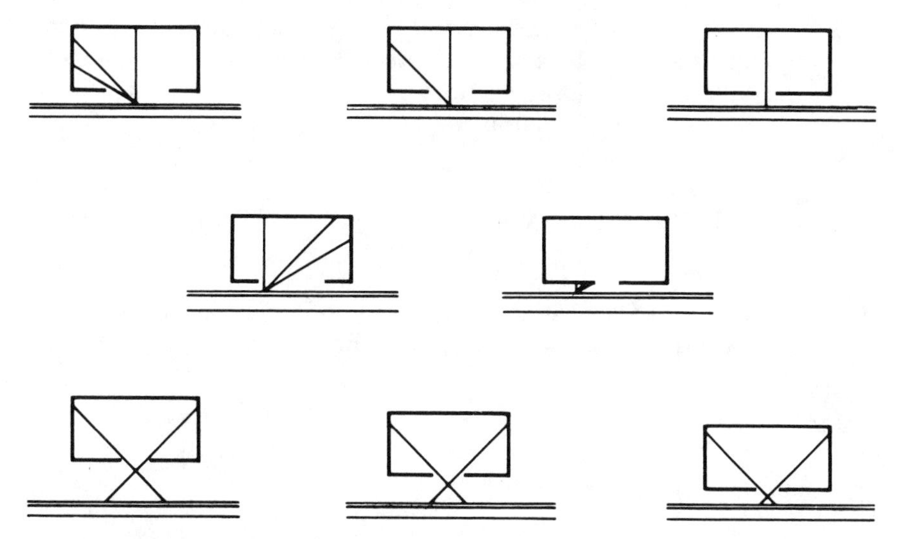

Figure 11.7. Diagrams showing how size of collimator openings can effect resolution.

As the detector array moves, counts at each point are directed to a corresponding memory location in an internal computer where they are stored for subsequent reconstruction of the activity pattern over the entire area scanned. The collimators have openings as small as 1×1 mm and as large as 10×10 mm. Openings are square for 2-D scanning while slits are best for 1-D work. Minimum dwell time per step is 0.001 min and the maximum is 10.0 min.

To locate activity sites without quantitation, using one of the larger collimators an entire 20 × 20-cm TLC plate can be scanned in under a minute; for the ultimate in resolution, that same plate might be scanned with a 1 × 1-mm collimator and at 1 min per point the scan would take $2\frac{1}{2}$ days and is too long to be useful. However, with efficiency for Geiger counting likely to be more than double that for proportional counting when so many of the events are rejected to obtain positional information, relatively fast scans with the Vanguard, perhaps on the order of 0.1 min per point, are practical and provide accurate information in a few hours or even less.

An entire 20 × 20-cm TLC plate scanned in 1-mm steps in both the *X* and *Y* directions gives 40,000 data points. Since 10 points are counted at a time, 4000 locations are examined, yielding a 2-D scan in which *X*- and *Y*- axis resolutions are the same. But, it may not be necessary to make so thorough an examination. To gain time, limits may be imposed upon the area of the scan and if 1-D tracks are being measured there is no requirement that *X*- and *Y*-direction resolutions be the same; therefore, slit collimators can be used. With 10 tracks on a plate, and with a 1 × 8-mm or 1 × 10-mm slit collimator, 200 or fewer locations need be measured and scanning times measured in minutes are the norm.

SCANNING THE AMERSHAM TEST PLATE WITH THE VANGUARD SCANNER

The Amersham Test Plate has been subjected to 1-D and 2-D scanning with the Vanguard instrument (Figures 11.8*a*–*g*). For 1-D scanning the tracks

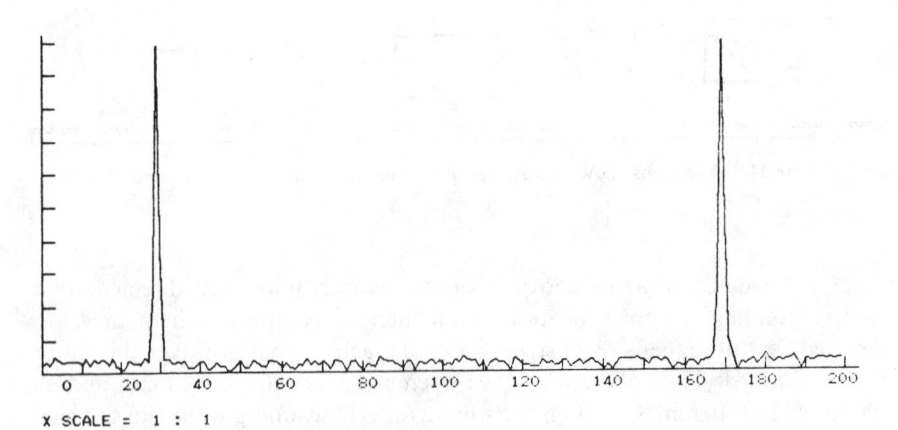

Figure 11.8a. Result showing the effect of collimator slit and resolution. Compare with Figure 11.3a.

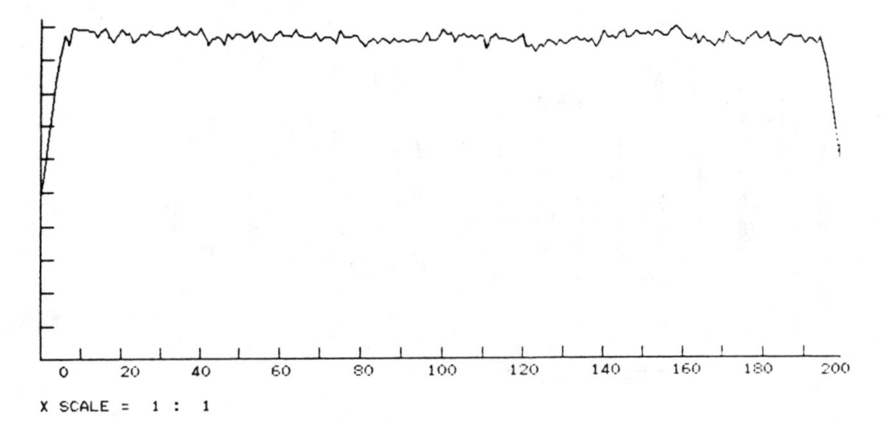

Figure 11.8b. Results of uniform field test.

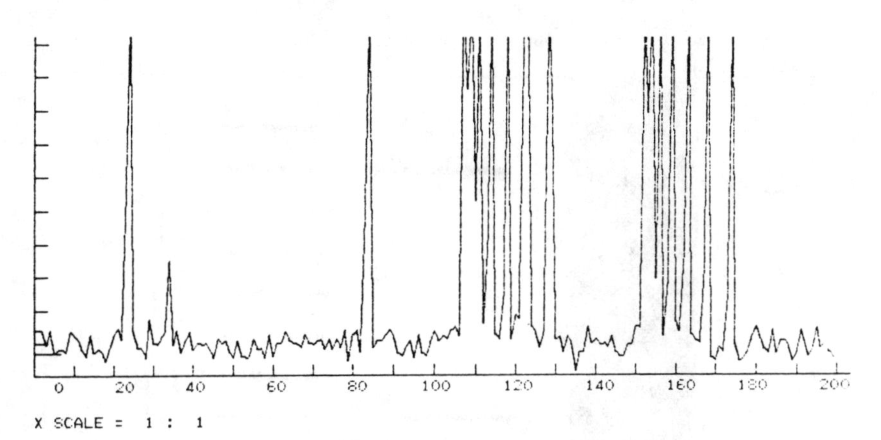

Figure 11.8c. Test for resolution of closely spaced strips with different count rates.

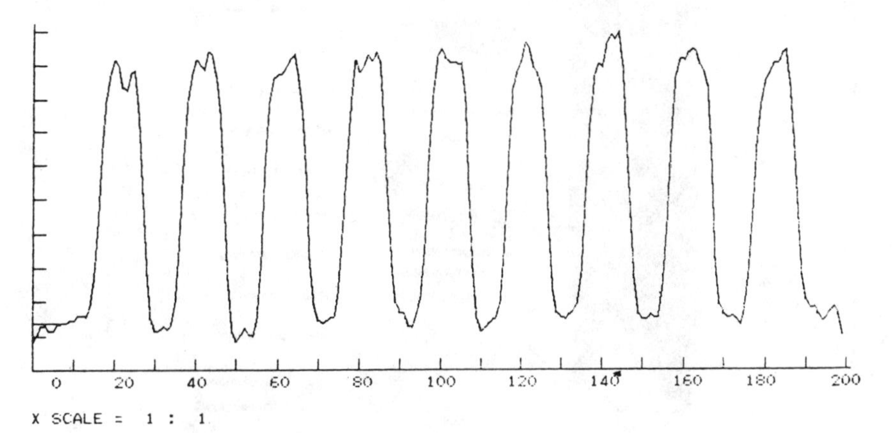

Figure 11.8d. Resolution is shown for 18 evenly spaced strips.

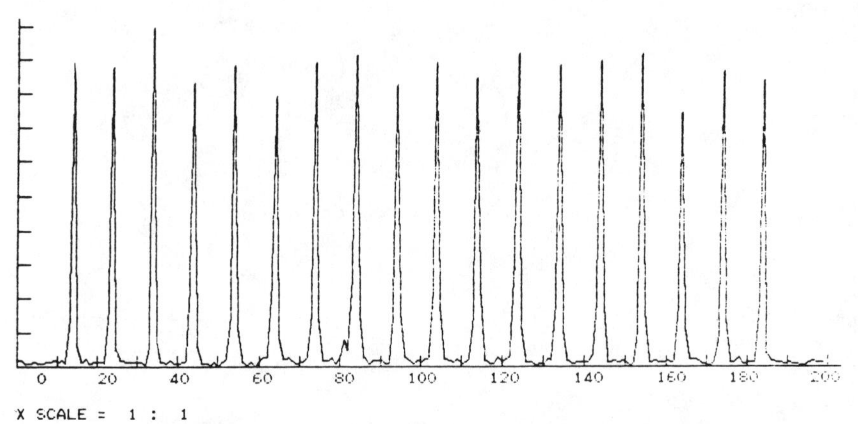

Figure 11.8e. Results of scanning nonuniform distributions about the centers of the track.

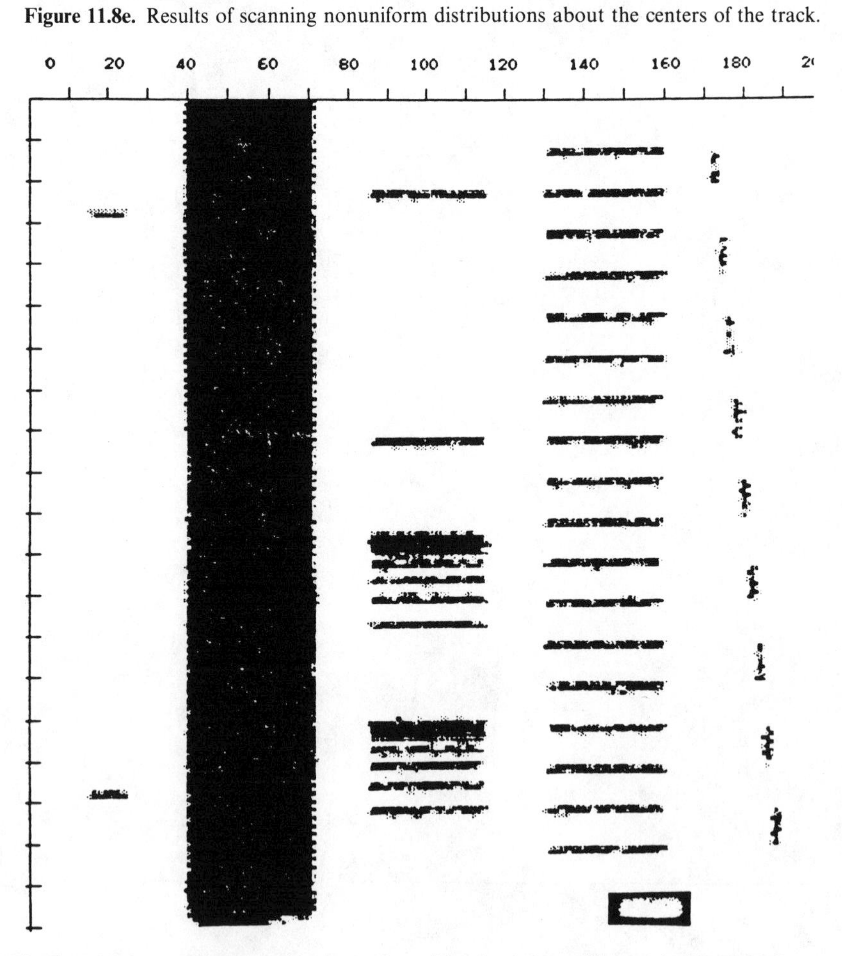

Figure 11.8f. Scan of the Amershlam test plate with a 1 × 1 mm collimator in 2D mode. See text.

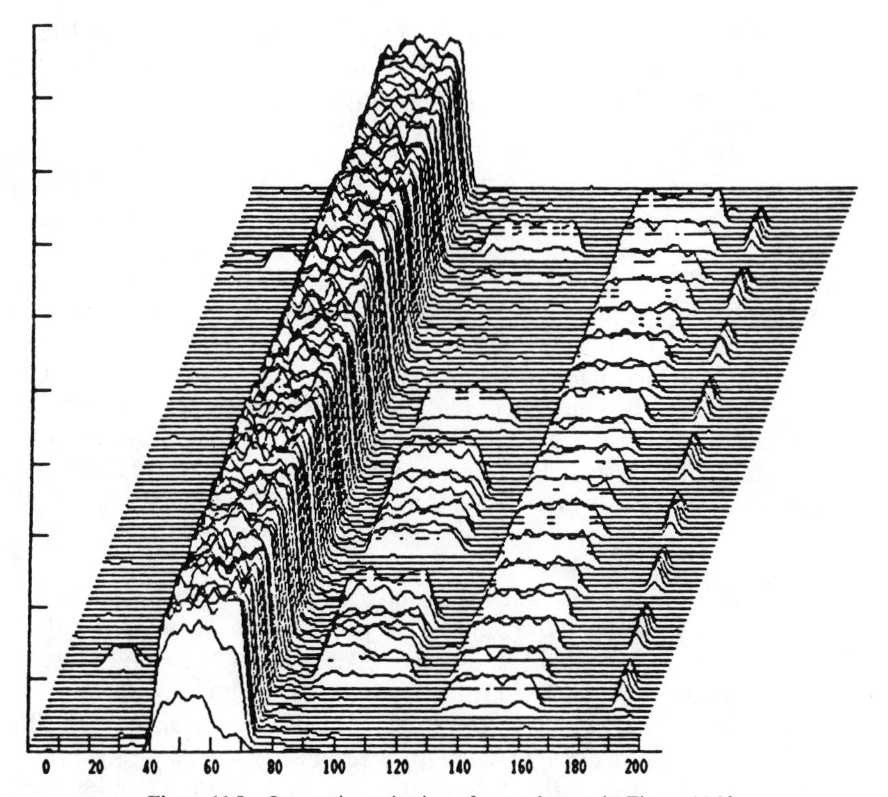

Figure 11.8g. Isometric projection of same data as in Figure 11.8f.

were oriented in the *Y* direction, a 0.5 × 8-mm slit collimator was used with the collimator slits perpendicular to and centered over the track, and the detector array was indexed in 1-mm steps in the direction of the tracks. Each of the detectors examined a tenth of the track; a 20-min scan of each track was made and the final record represents the composite from all 10 detectors. Upon completion of each pass, the array was programmed to move over the center of the next track and another 20-min scan was made.

The 2-D scan was made with a 1 × 1-mm collimator, the detector array being stepped in 1-mm increments in each direction. Elapsed time for the complete scan, including that required for mechanical motion, was approximately 20 h, corresponding to a dwell time per point of 0.25 min.

Figure 11.8a: Track 1 (20 mm). The two activity strips at 30 and 170 mm are found in their proper locations. Each peak appears to be about 4 mm wide at baseline, some of which can be attributed to the small probability that the center of the collimator slit has positioned exactly over the center of the strip. Any such dissymmetry would perforce widen the peak by 1 mm at baseline; the remainder of the widening is primarily related to the distance between the surface of the sample and the bottom of the collimator.

Figure 11.8b: Track 2 (55 mm). The Vanguard came out very well in the uniform field test, the maximum deviation being less than $\pm 3\%$ of the mean.

Figure 11.8c: Track 3 (100 mm). A severe test of the PSPC combines measurement of the resolution of closely spaced strips with response to different count rates. In each of the two seven-strip clusters all strips have the same activity and all seven peaks can be clearly seen, even those spaced only 0.5 mm apart. With 0.5-mm separation, the trough between peaks is 50% of peak height, with 1-mm separation the trough is only 30%; near baseline separation is not achieved until the strips are at least 2 mm apart. This is in contrast to the PSPC which was unable to see the 0.5-mm separation, could barely discern 1-mm separation, and gave a trough of only 70% for 2-mm separation.

With closely spaced strips differing in activity by factors of 1:10:100:1000, in the data presentation shown, the small peaks cannot be seen and the peakfinder apparently could not locate them. This illustrates the importance of scale expansion, not employed here, but, in fact, a part of the software of the Vanguard as well as all of the PSPCs. Even if the system counted perfectly, with the form of data presentation used here where the largest peak is not quite full scale, peaks one-hundredth or one one-thousandth the amplitude of the reference would never be seen.

Figure 11.8d: Track 4 (145 mm). The Vanguard performed exceptionally well on the 18 evenly spaced strips. Baseline resolution is attained; integration showed the peak contents to be quite uniform.

Figure 11.8e: Track 5 (180 mm). This test of performance for activity not uniformly distributed about the center of the track simulates the potential measurement problem that might occur if tracks do not run straight. The nine bars spread the activity over 10 mm of width. Each should give the same count rate and show a flat top. With the Vanguard all peaks are measured and their activities are comparable but the hoped for flat-topping seems to be lost in the statistics.

Figure 11.8f: 2-D Scan (Topographic Presentation). A 1:1 printout of the 2-D scan of the Amersham test plate with a 1×1-mm collimator provides a picture of activity location. With this test plate, the gray scale of eight is not as helpful as it otherwise might be since all peaks that are detected have the same activity. However, it is obvious from this presentation and that below that the uniform activity field has a greater activity density than do the activity strips. In the illustration, background has been subtracted and the regions between the peaks are clean. The measured width of the individual 1-mm strips is 3 mm, confirming the results of the 1-D scan discussed above. Overlap of the 0.5- and 1-mm spaced strips in track 3 results from the failure to achieve baseline separation and illustrates the need to achieve the best possible resolution to properly interpret results.

Figure 11.8g: 2-D Scan (Isometric Presentation). An isometric projection of the same data used for the topographic scan shows how the two 2-D presentations complement one another. The topographic picture accurately shows location while the isometric is more informative of relative activity levels. But neither one, nor the two together, provides the complete picture. That comes only when they are combined with some means for obtaining digital information, either sequential 1-D printouts of various lanes with a peakfinder providing integrals, or one of several methods for encircling a peak on a computer screen and integrating the counts within the chosen boundaries.

SCANNING ^{32}P SAMPLES WITH THE VANGUARD SCANNER

With the Vanguard scanner there is almost nothing different about scanning ^{32}P samples than would be obtained if the samples contained ^{3}H or ^{14}C, one concession to the higher energy being the use of a double-thickness collimator. Also, when maximum resolution, that is, 1 mm, is sought it has been found advantageous to employ a 0.5-mm collimator—for 1-D it is a slit collimator with ten 0.5×8-mm openings, while for 2-D the collimator has 10 circular openings equivalent in area to 0.5×0.5 mm. In either case the stepping increment is 1 mm, with the undersized openings serving to minimize the effects of angular radiation components which, nevertheless, do find their way into the detector and are counted.

What can be achieved with ^{32}P is illustrated in the following discussion. An autoradiograph of a TLC plate containing six tracks with labeled phosphorous-containing insecticides together with the topographic and isometric projections from a scan of the same plate with the Vanguard instrument indicated that the autoradiograph does not lend itself to quantitation. It is inherently nonlinear and once it has been fogged to its maximum density there is not even the possibility of guessing as to activity level. And, with the relatively short half-life of ^{32}P (14 days), making a new exposure after an initial one has been found to be too long or too short is fraught with problems.

A 1:1 topographic projection accurately provides activity location but barely tells us anything about the levels. It does suggest that the peak at the end of each of the two center tracks might have more than one component, something that one also might guess from the autoradiograph. On the other hand, this is an easy call when the isometric projection is examined, the two components showing up clearly even though the centers of the two zones are less than 5 mm apart and the principal component is present at many times the activity level of the other.

SCANNING ^{99m}TC WITH THE VANGUARD SCANNER

There is increasing interest in scanning TLC plates and paper strips for "Tc-99" for control of preparations used in medical imaging. But in discussing Tc-99 it is necessary to be more explicit. Tc-99 is a beta emitter with an E_{max} of 292 keV and a half-life of 2×10^5 years. Tc-99m is primarily a gamma emitter with an energy of 141 keV and a half-life of 6 h. It is most often obtained in the laboratory by "milking" a Mo-99 "cow". The cow, an ion-exchange column, is loaded with longer lived Mo-99 which continually decays to Tc-99m. This, in turn, can be washed off the column with suitable elutants. Then, because of the short half-life of Tc-99m, whatever chemistry and testing that is performed must be done quickly.

The principal use for Tc-99m is in medical imaging where its short half-life and low-energy gamma emission make it safe and useful. But, because of the short half-life of Tc-99m, Tc-99 is often used for the pilot studies—chemical synthesis, separation, and purification—which must precede clinical use of Tc-99m. An analytical procedure ultimately intended to test Tc-99m preparations should be equally applicable to that same preparation containing Tc-99. TLC and paper chromatograph scanning is one obvious technique that meets this qualification.

Scanning Tc-99 with the Vanguard is like scanning any other beta emitter. Being twice as energetic as C-14 ($E_{max} = 156$ keV) but only about a fifth as energetic as P-32, one should expect and does obtain 1-mm resolution, good counting efficiency, and, because those who work with Tc-99 almost always have a great deal of it, short scans. Resolution is significantly better than that which can be had with a position-sensitive proportional counter because PSPC performance has already begun to degrade at the C-14 energy level and an isotope twice as energetic will exhibit even greater performance degradation.

What about scanning Tc-99m? With a 6-h half-life the chromatography must be fast and the scan must give good resolution quickly. Good resolution is important; impurities often are not well separated from the main component.

The problem is that counting gas, proportional or Geiger, is not really able to stop gammas. But there is a method suitable for Tc-99m and other low-energy gamma emitters and it is providing good results. Collimators have been developed that are constructed of two standard thicknesses of steel which sandwich a comparable thickness of lead. This shielding is sufficient to keep a low-energy gammas from entering the counting chambers except at the collimator opening. And at the inside of each opening there is a fine-mesh copper screen whose purpose is to intercept gammas and create secondary electrons. These then count as do betas.

Printouts of Tc-99m scans show what can be done with one of the new collimators. But, before they are discussed, it is useful to consider the situation up to now. If a standard-thickness collimator is used to scan a test plate containing closely spaced spots of gamma activity, some activity will penetrate the collimator and cause counts even before the opening is over a spot. Further, some of the gammas that are stopped by the collimator generate secondary electrons within the counting chamber, most of which will be counted. The Vanguard system concept presumes that the only activity that is counted is directly under the collimator opening, but with a thin collimator that is obviously not the case; the net result is that closely spaced spots cannot be resolved.

^{99m}Tc test samples were scanned with the Vanguard scanner to compare standard and special collimators. Results are given for four plates. Each is 80 mm long and they were positioned along two tracks ($Y = 20$–21 cm,

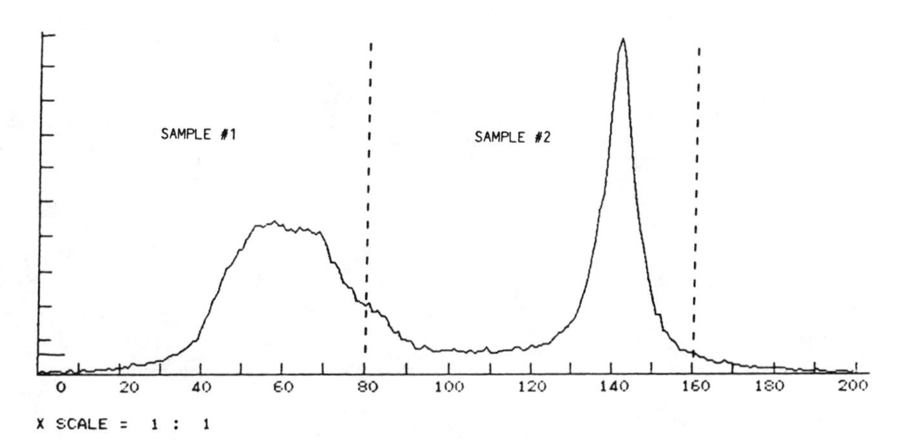

Figure 11.9. Scan of a sample using an 8×0.5 mm collimator.

$Y = 50$–51 cm) so that each trace provides results for two plates, one from 0 to 80 mm and the other from 80 to 160 mm. To facilitate examination, dashed lines are drawn at the limits of the samples. A standard Vanguard collimator 8×0.5 mm was used to obtain the data of Figures 11.9 and 11.10 while a 10×1-mm double-thickness steel collimator with a copper screen (the triple-thickness collimator was not then available) gave the data shown on Figures 11.11–11.16. Brief comments on the individual figures might be helpful:

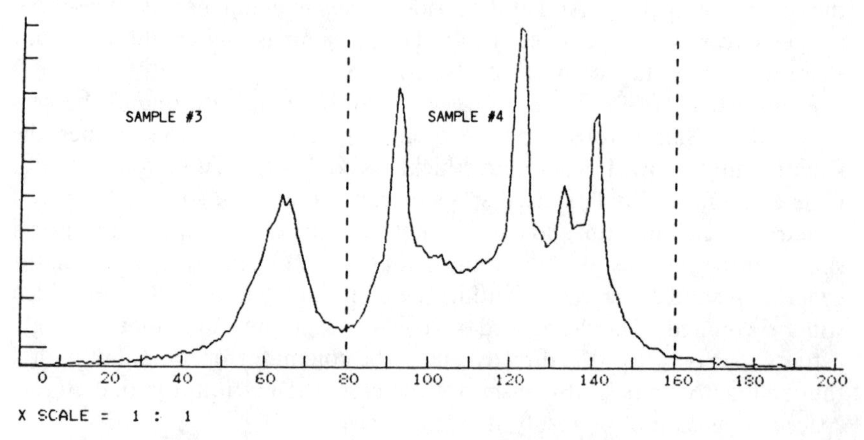

Figure 11.10. The standard collimator suggests that sample 3 is a single peak. Sample 4 appears to be several peaks. These conclusions may be in error. See subsequent figures and text.

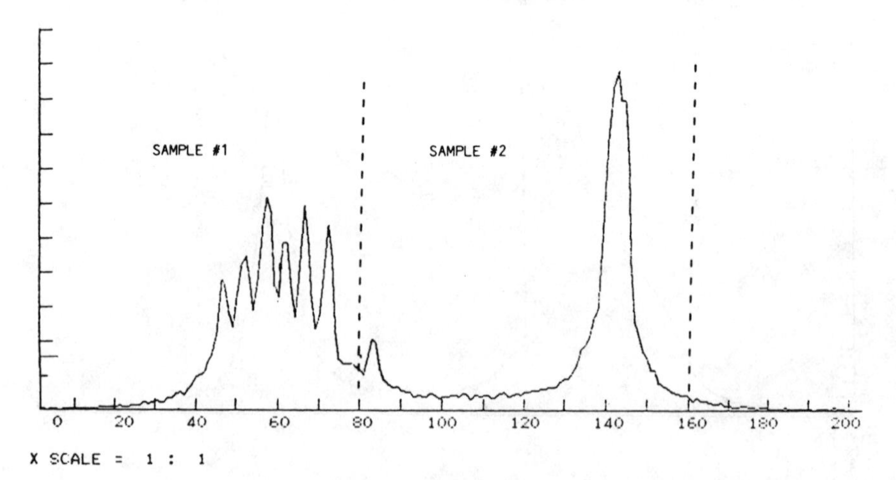

Figure 11.11. Sample/scanned with a copper mesh collimator. See Figure 11.1.

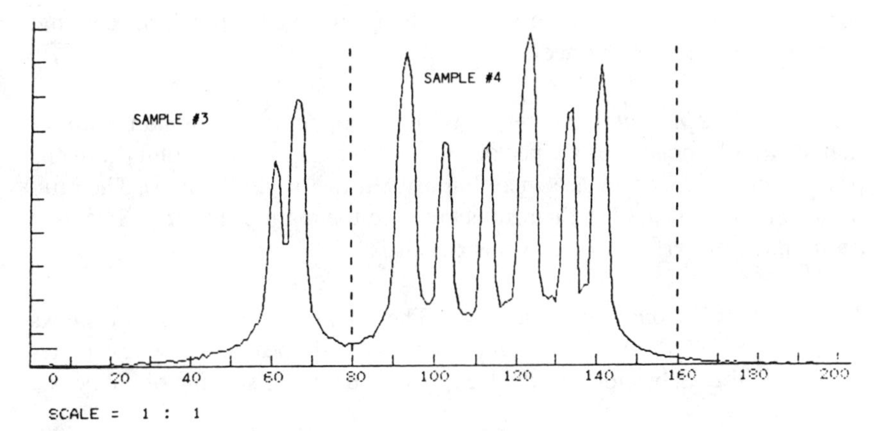

Figure 11.12. With the copper mesh collimator, Sample 3 is in fact two peaks. Sample 4 has resolved to six peaks. See Figure 11.10.

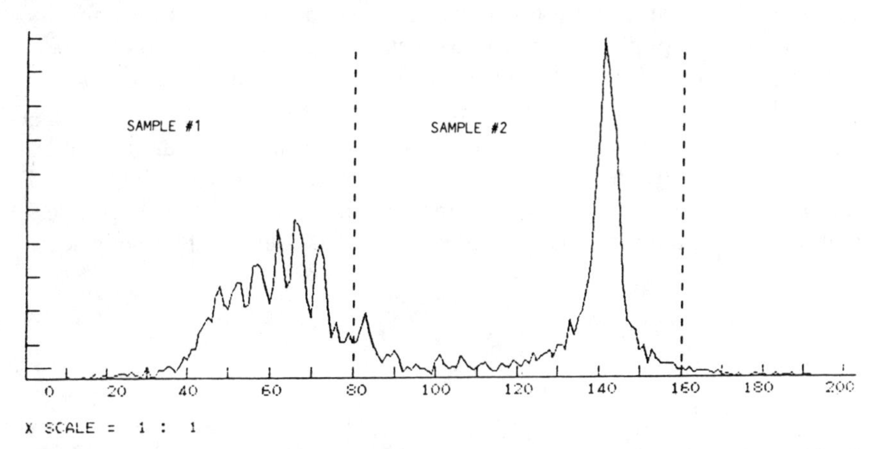

Figure 11.13. Compare these traces to those of Figure 11.11. This shows the importance of dwell time.

Figure 11.9 (2 min total scan time). A scan made with the standard collimator suggests that samples 1 and 2 each consists of a single broad peak, with that of sample 2 being better defined. In further discussion we will see that this is an erroneous picture.

Figure 11.10 (2 min total scan time). The standard collimator suggests that sample 3 has but a single peak. Sample 4, however, does appear to be composed of several peaks, possibly totaling four. Again, it will be seen that both conclusions are erroneous.

Figure 11.11 (2 min total scan time). With the copper mesh collimator we see that sample 1 consists of six peaks. We also see improved resolution of the peak in sample 2 as well as a small impurity at the origin (83 mm). The ratio of the impurity (peak 7 in the results table) to the main peak (peak 8) is used as an index of performance; in this case it is 9.3%.

Figure 11.12 (2 min total scan time). The copper mesh collimator shows sample 3 to be a doublet and sample 4 to have six peaks. These test plates were all deliberately spotted, which explains the regular spacing of the peaks.

Figure 11.13 (0.2 min total scan time). The traces show the same characteristics as those of Figure 11.15. However, the attempt to scan the samples so rapidly did not produce as good results and it was concluded that the dwell time of 0.01 min per point was insufficient. Note, however, that the impurity at the origin of sample 2 (83 mm) was seen and that the 7:8 ratio of 11% is not very different from that obtained with a ten times longer scan.

Figure 11.14 (0.2 min total scan time). Here, too, the 0.2-min scan of samples 3 and 4 is not really adequate. Nevertheless, sample 3 is seen as a doublet and sample 4 does show six peaks, with the second, third, and fifth having less amplitude than the first, fourth, and sixth, as was found in the 2-min scan.

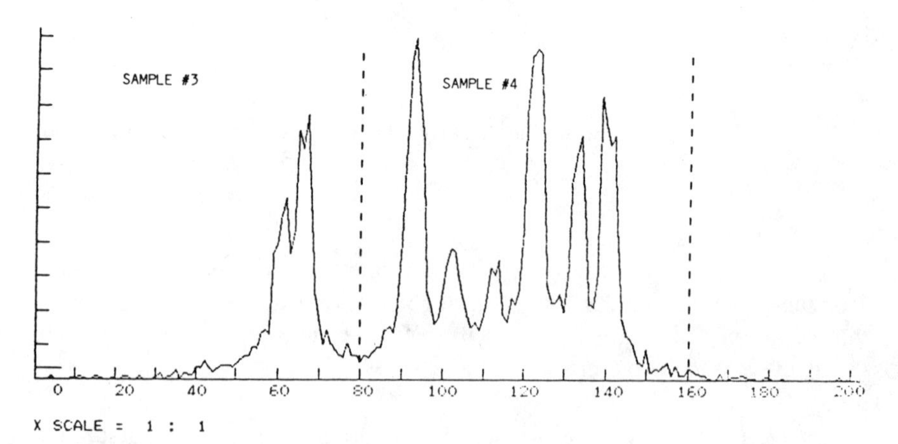

Figure 11.14. Comparison with Figure 11.12 shows the effect of a 0.2 min scan time.

Figure 11.15 (4 min total scan time). Increasing the scan time from 2 to 4 min did not provide any further enhancement.

Figure 11.16 (4 min total scan time). Again, the added time provides no advantage over the 2-min scan.

This data clearly shows the Vanguard to be a useful device for ^{99m}Tc scanning.

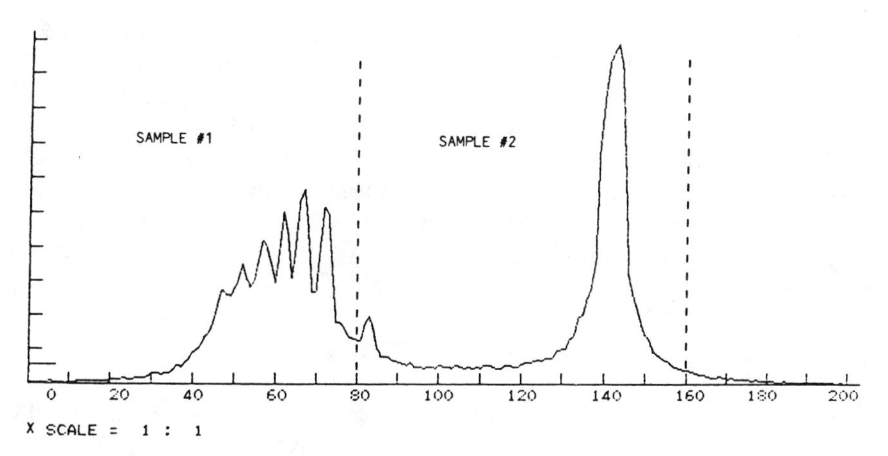

Figure 11.15. Increase of scan time from 2 to 4 min shows no advantage.

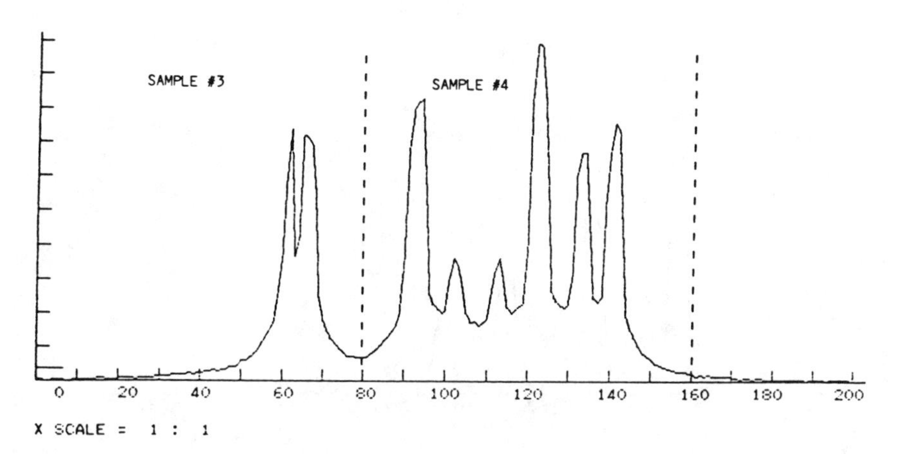

Figure 11.16. Added time again shows no advantage. See Figure 11.14.

SCANNING ^{111}IN WITH THE VANGUARD SCANNER

^{111}In is a gamma emitter with two principal emissions, 171 and 242 keV, and with a 2.81-day half-life. The outcome of scanning a labeled test plate is found in Figure 11.17. Eight levels of activity over a 100-fold range in the indicated ratios were spotted on approximately 20-mm centers in each of two lanes; numerical results are shown for one:

Spot No.	1	2	3	4	5	6	7	8
Applied	10.0	9.0	7.0	5.0	3.0	1.0	0.5	0.1
Found	10.0	8.3	7.2	4.9	2.6	0.9	0.4	0.2

A triple-thickness slit collimator (1 × 10 mm) with a copper mesh target was employed; the detector array was stepped 1 mm down the length of the lane after each 0.5-min counting period. Near baseline resolution was achieved; apparent peak width at baseline was approximately 20 mm for the most active peaks, but ony 5–10 mm for the less active. The gamma activity appeared to create no problems.

THE VANGUARD SCANNER IS A USEFUL DEVICE

The Vanguard scanner has proven to be a useful device for 1-D and 2-D scanning of TLC plates and other planar samples. Primarily intended to replace position-sensitive proportional counters for the measurement of more energetic beta emitters and soft gamma emitters, it represents a

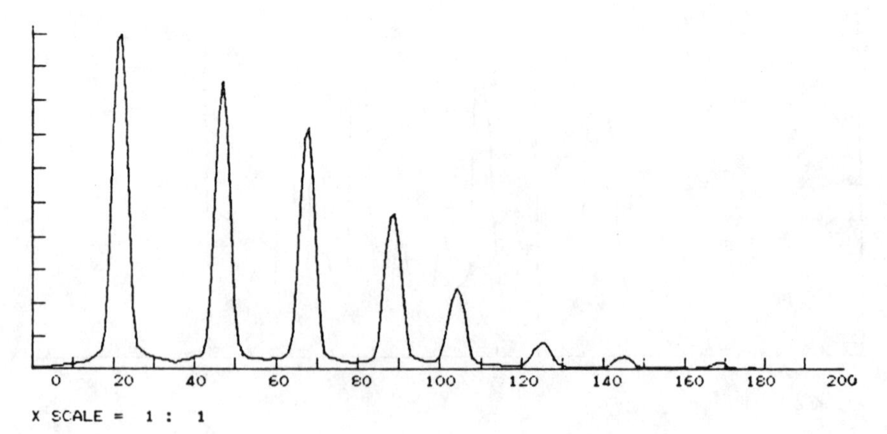

Figure 11.17. Scan of ^{111}.In on a test plate with eight levels of activity over a 100-fold range.

successful return to proven principles. The Vanguard scanner does not incorporate any breakthroughs or novel implementations. Rather it is based upon the application of modern technology to methods that were perfected long ago. That has been coupled with the recognition that electronic component costs have continually declined over the years, thereby allowing multiple parallel measurement with consequent saving of operating time.

REFERENCES

1. C. J. Borkowski and M. K. Kopp, Position–Sensitive Radiation Detector. U.S. Patent 3,483,377 (1969).
2. G. Charpak, R. Bouclier, T. Bressani, J. Favier, and C. Zupancic, *Nucl. Instrum. Methods* 62: 262 (1968).
3. C. J. Borkowski and M. K. Kopp, *IEEE Trans. Nucl. Sci.* NS 17: 340 (1970).
4. C. J. Borkowski and M.K. Kopp, *IEEE Trans. Nucl. Sci.* NS 193: 161 (1972).
5. A. Gabriel and S. Bram, *FEBS Lett.* 39: 307 (1974).

CHAPTER

12

BIOANALYTICAL APPLICATION OF THIN-LAYER CHROMATOGRAPHY/FOURIER TRANSFORM INFRARED SPECTROMETRY

JAMES A. HERMAN AND KENNETH H. SHAFER

The benefit of infrared (IR) detection used in thin-layer chromatography (TLC), as with any information-rich detector, is the capability to identify separated components with spectral rather than chromatographic information. Besides achieving higher confidence levels in identifications by TLC/FT-IR compared to TLC alone, it is advantageous not to have to maintain authentic standards. The sample spectrum need only be compared to its reference spectrum for identification.

Thin-layer chromatography complements infrared analysis in that separation of mixtures of nonvolatile or thermally labile compounds can be performed utilizing a relatively simple chromatographic technique. Indeed, the effort involved in separation of most samples by TLC is comparable to that required for conventional methods of sample preparation in infrared spectroscopy.

The direct measurement of TLC fractions on the plate can be performed by diffuse reflectance or photoacoustic spectrometry (1–4). However, in situ measurement can result in band shifts due to intermolecular interactions that occur between the analyte and stationary phase. The spectrum of the analyte will then appear different than the reference spectrum of the same compound obtained using conventional methods of sample preparation. The intensity of analyte bands are also strongly attenuated in regions where the silanol groups of the stationary phase absorb. The alternative to in situ measurement is to transfer the TLC fractions to an IR transparent substrate prior to IR measurement (5–9). Although the transfer approach is less straightforward than the in situ measurement, better quality spectra are obtained.

The commercial accessory is based upon a sample transfer approach by Shafer et al. (10) in which the mobile phase is made to flow sequentially in two directions as in two-dimensional TLC. The separation is performed in the first direction, followed by transfer in the second direction. The TLC fractions are thereby transferred simultaneously to an IR transparent powder for diffuse reflectance measurement.

Bioanalytical applications often pertain to the analysis of thermally labile or nonvolatile samples. The pharmaceutical product methyl prednisolone acetate is such a sample chosen for analysis by TLC/FT-IR. The corticosteroids tend to be thermally labile and nonvolatile. Because infrared spectrometry is a nondestructive sample technique, fractions can be analyzed further. In this paper, the TLC fraction identified as methyl prednisolone acetate by infrared spectrometry is then analyzed by fast atom bonbardment mass spectrometry (FABS).

TLC has been shown to be one of the most effective and versatile techniques for the separation of complex lipids (11). In this paper, a standard mixture of phospholipids are analyzed to show the identification capability of TLC/FT-IR.

EXPERIMENTAL

TLC/FT-IR Procedure

Analysis begins with a conventional separation on a TLC plate. The plate is spotted with a sample and placed in the development chamber. The sample elutes up the plate with individual components separating along the way. When the separation is complete, as shown in Figure 12.1, the plate is

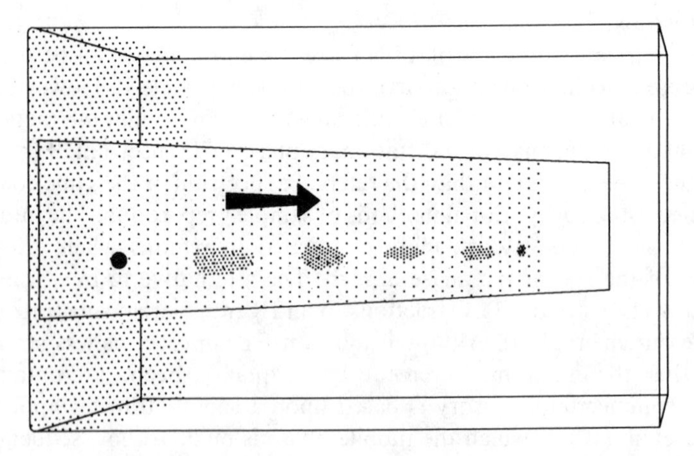

Figure 12.1. Conventional TLC separation. components are separated up the length of the plate using standard TLC methodology.

removed, dried, and fastened to the Optitrain, as shown in Figure 12.2. The Optitrain and plate are then placed into the unitary transfer assembly as shown in Figure 12.3. Solvent is injected in controlled amounts, and the sample moves off the plate into a row of stainless steel wicks in the Optitrain. At the top of each wick is a cup filled with an IR transparent diffuse reflectance powder as shown in Figure 12.4. A controlled flow of air is directed across the surface of the Optitrain, drawing the separated

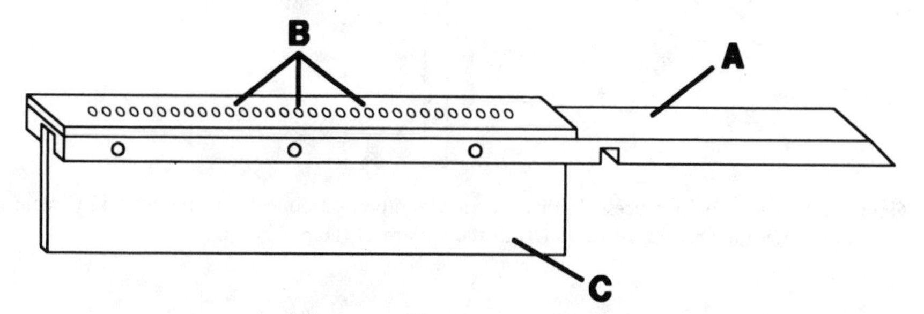

Figure 12.2. Optitrain and TLC plate. The Optitrain (*A*) contains a row of sample cups (*B*) to hold separated components. The TLC plate (*C*) is attached to the underside prior to sample transfer.

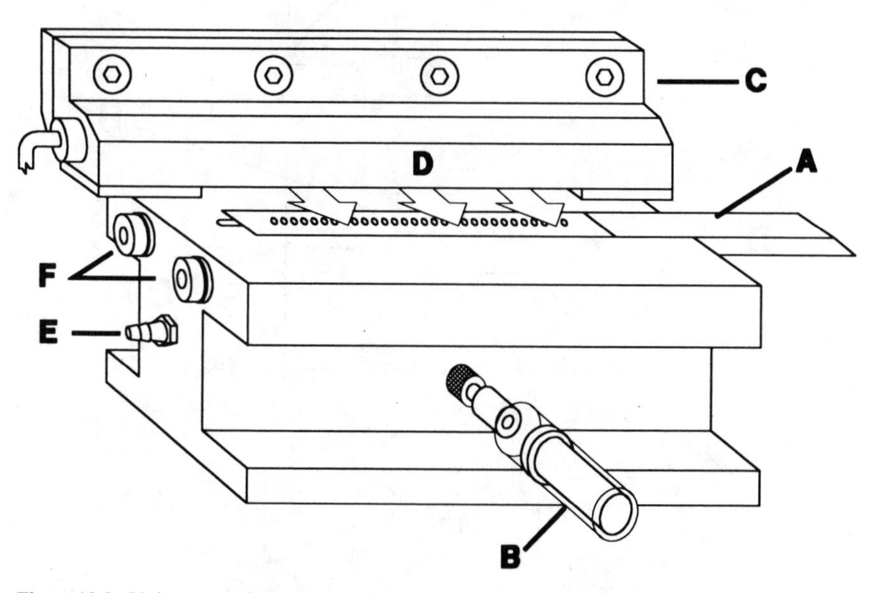

Figure 12.3. Unitary transfer assembly. Shown are the optitrain (*A*), the solvent injection syringe (*B*), the air knife (*C*), and airflow over the sample cups (*D*), solvent overflow relief (*E*), and connections for water bath circulation (*F*).

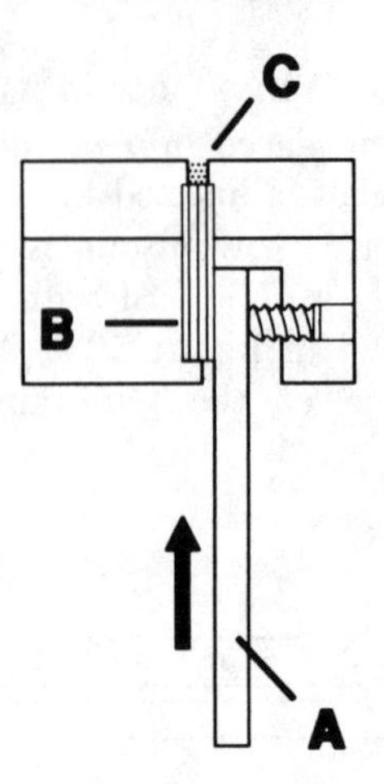

Figure 12.4. The transfer process. In this end view, sample components move up the TLC plate (*A*) and though the Optitrain's wicks (*B*) to the surface of the powder (*C*).

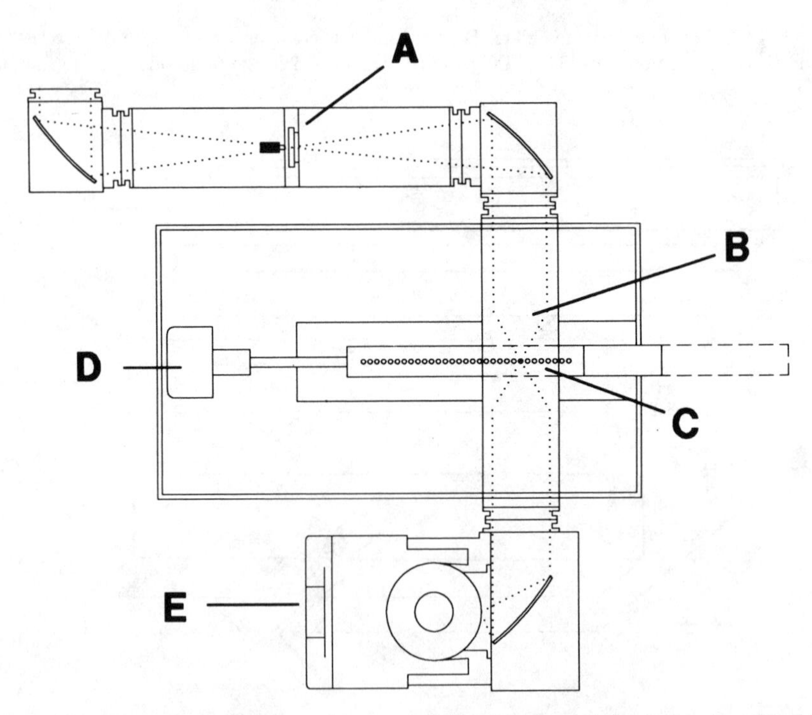

Figure 12.5. The automated diffuse reflectance sampler. This top view shows the adjustable field stop (*A*), matched set of paraboloids (*B*), Optitrain cups (*C*), stepper motor (*D*), and the MCT detector module (*E*).

components up through the wicks and powder until they are concentrated on top of their respective cups. The entire transfer is accomplished in a matter of minutes. Next, the Optitrain (with separated components concentrated on top of its sample cups) is inserted into the automated diffuse reflectance sampler, as shown in Figure 12.5. The infrared beam from the spectrometer is focused onto the sample cups and the diffuse reflected light is collected by a matched set of paraboloids. Individual cups are positioned under the beam by a precision stepper motor as each component is analyzed. Sample positioning and data acquisition are controlled by computer. Sample information consists of a reconstructed chromatogram representing the contents of the Optitrain, individual spectra of each component, and "best match" reports from spectral library searches. Infrared spectra were acquired on an Analect RFX-65 equipped with the Chromalect TLC/FT-IR peripheral. Spectra were obtained at a resolution of 4 cm^{-1}. Each chromatographic separation was analyzed in about 30 to 40 min.

Methyl Prednisolone Acetate Preparation

Analtech prescored 2.5 × 10-cm precoated silica gel plates were used after precleaning with acetone. Solutions were applied by micro cap deposition with subsequent drying with cool air from a forced-air gun for 2 min. The mobile phase consisted of chloroform, methanol, and water at 180:15:1 (v/v). A 1-mL volume of an aqueous solution containing 40 mg of methyl prednisolone acetate and phosphate buffer was extracted in 1 mL of dichloromethane. A volume of 1 μL was deposited on the plate for analysis. The solvent front was developed 10 cm. TLC fractions were transferred in 5 min with 10 mL of dichloromethane. A TLC fraction identified by infrared spectrometry as methyl prednisolone acetate was then extracted from the wicks using 5 μL of methanol and analyzed by FABS. The sample was recorded in thioglycerol. The analysis was performed at the mass spectrometry facility maintained at the University of California, Riverside.

Phospolipids Preparation

The following standard phospholipids were purchased from Sigma Chemical Company: phosphatidyl choline, egg (lecithin), phosphatidyl ethanolamine (egg), and phosphatidyl glycerol (synthetic). Standard solutions of the phopholipids were prepared in chloroform–methanol 1:1 (v/v) at a concentration of 1 mg/mL. Silica gel plates were washed with methanol and ammonium hydroxide (33%) at 75:25 (v/v). The mobile phase consisted of chloroform, methanol, and ammonium hydroxide (33%) at 65:25:4 (v/v). A 1 μL volume of sample was deposited on the plate. The solvent front was developed 10 cm.

TLC fractions were transferred in 10 min with 10 mL of chloroform and methanol at a 1:1 (v/v).

RESULTS AND DISCUSSION

The spectra of the TLC fraction and standard of methyl prednisolone acetate are almost identical, as shown in Figure 12.6. The infrared spectrum of the analyte is of the intact molecule. Successful analysis of the thermally labile

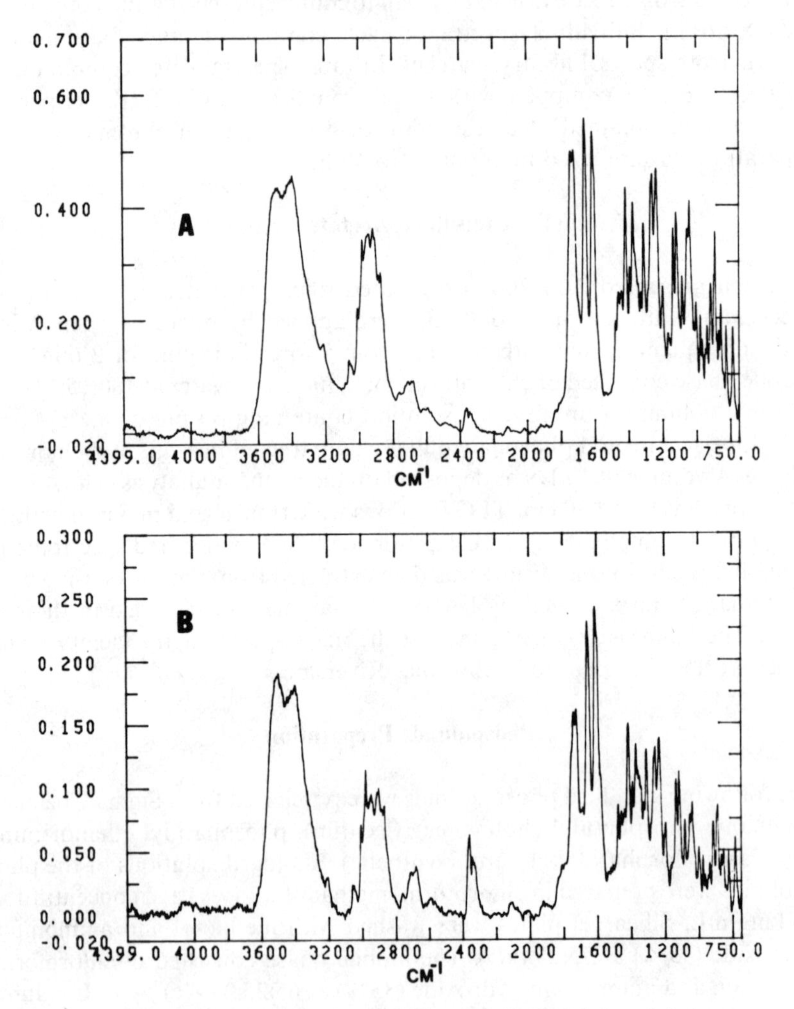

Figure 12.6. Spectra of (*A*) TLC fraction and (*B*) standard of methyl prednisolone acetate.

compound is possible because the solvent is eliminated without heat by using a flow of air. The mass spectrum of the TLC fraction analyzed by FABS is shown in Figure 12.7. The spectrum shows a pseudomolecular ion at 417 characteristic of the $[M \times H]^+$ ion (12). Peaks at 181, 217, and 257 are from the matrix of thioglycerol. The TLC fraction is quite easily analyzed by

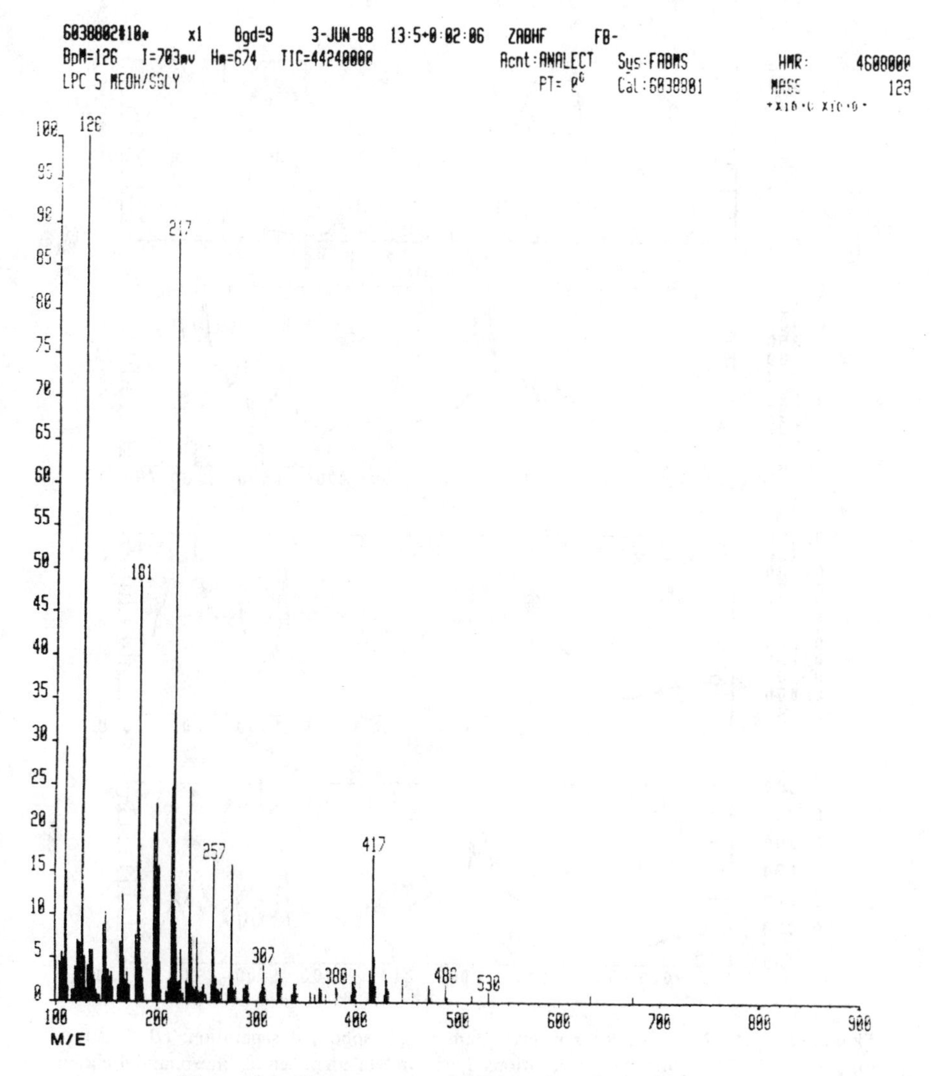

Figure 12.7. Mass spectrum of TLC fraction obtained by FABS.

FABS because it is isolated during the TLC/FT-IR analysis onto an inert substrate in a concentrated amount.

Although the separation of phosphatidyl choline and phophatidyl ethanolamine is only partial, as indicated by the reconstructed chromatogram shown in Figure 12.8*A*, identification is still possible by analyzing those portions of the spots that do not overlap. The sample spectra obtained at cup locations 9, 11, and 16 were identified as phosphatidyl choline, phosphatidyl ethanola-

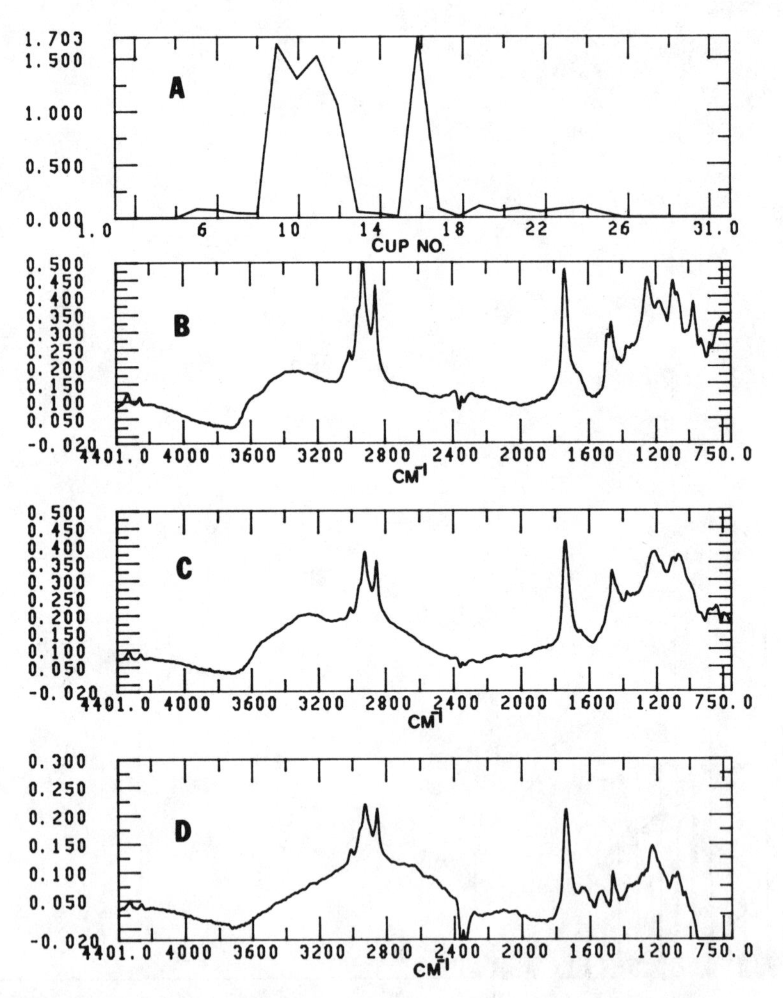

Figure 12.8. (*A*) Reconstructed chromatogram of phospholipid separation; (*B–D*) spectra obtained of TLC fractions at cup locations 9, 16, and 11 identified as phosphatidyl choline, phosphatidyl glyerol, and phosphastidyl ethanolamine, respectively.

mine, and phosphatidyl gylcerol, respectively. Sample spectra of the three phospholipids are shown in Figures 12.8*B–D*. Each sample spectrum compares well to the reference spectrum obtained by diffuse reflectance measurement of the standard. An example of the good quality matches is shown in Figure 12.9 for phosphatidyl ethanolamine.

Applications must be limited to the separation of samples containing a relatively small number of compounds for successful identification. If this is not possible, structural information can still be obtained regarding class

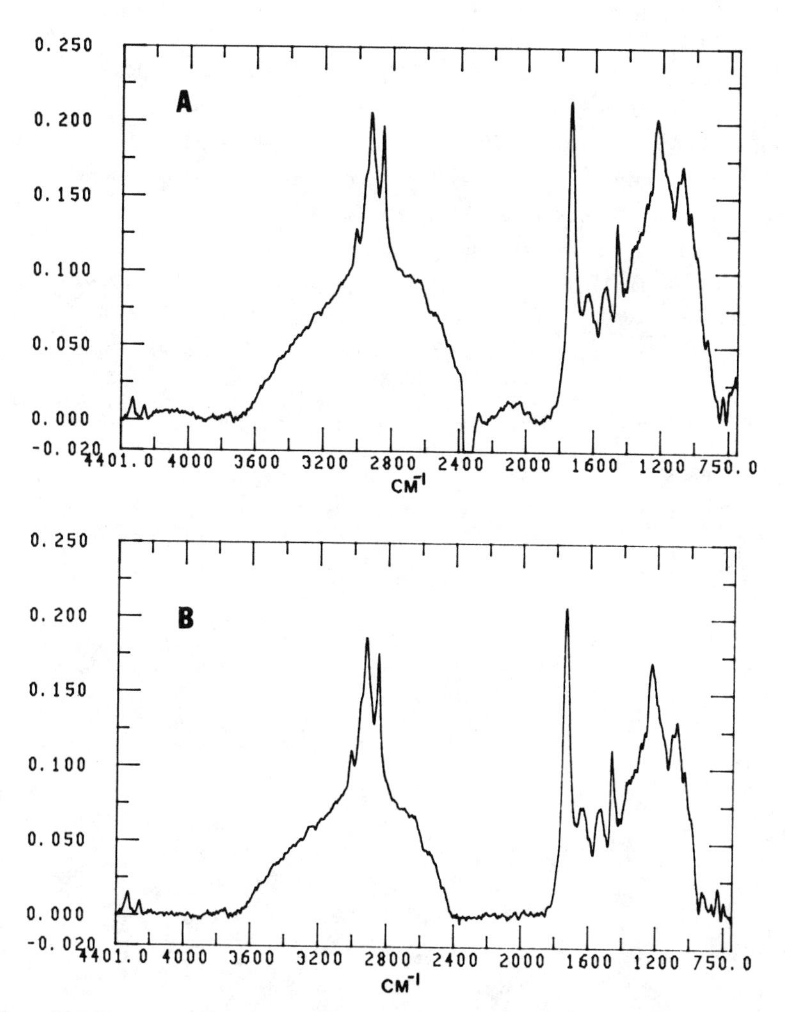

Figure 12.9. Spectra of (*A*) TLC fraction and (*B*) standard of phosphatidyl ethanolamine.

separations for mixtures that are more complex. The sample transfer approach described in this paper provides a practical and accurate means for achieving infrared identification of the separated components of a nonvolatile or thermally labile sample.

REFERENCES

1. M. P. Fuller, and P. R. Griffiths, *Anal. Chem.* 50: 1906 (1978).
2. G. E. Zuber, R. J. Warren, P. P. Begosh, and E. L. O'Donnell, *Anal. Chem.* 56: 2935 (1984).
3. L. B. Lloyd, R. C. Yeates, and E. M. Eyring, *Anal. Chem.* 54: 549 (1982).
4. R. L. White, *Anal. Chem.* 57: 1819 (1985).
5. M. P. Fuller and P. R. Griffiths, *Appl. Spectrosc.* 34: 533 (1980).
6. H. R. Garner and H. Packer, *Appl. Spectrosc.* 33: 1323 (1967).
7. K. O. Alt and G. Szekely, *J. Chromatogr.* 202: 151 (1980).
8. J. M. Chalmers and M. W. Mackenzie, *Appl. Spectrosc.* 39: 634 (1985).
9. J. M. Chalmers, M. W. Mackenzie, and J. L. Sharp, *Anal. Chem.* 59: 415 (1987).
10. K. H. Shafer, P. R. Griffiths, and W. Shu-qin, *Anal. Chem.* 58: 2708 (1986).
11. M. Kates, *Techniques of Lipidology-Isolation, Analysis, and Identification of Lipids,* 2d ed. Elsevier, New York, 1986.
12. J. Belanger, B. A. Lodge, J. R. J. Pare, and P. Lafontaine, *J. Pharm. Biomed. Anal.* 3: 81 (1985).

CHAPTER

13

DETECTION OF RADIOACTIVITY DISTRIBUTION WITH POSITION-SENSITIVE DETECTORS, LINEAR ANALYZER, AND DIGITAL AUTORADIOGRAPH

HEINZ FILTHUTH

What do we want to measure? Position and intensity of surface distributions of ionizing radiation, especially from TLC plates with labeled zones, electrophoresis gels, protein distributions, DNA sequences, blots, and tissue sections are considerations. Figure 13.1 shows the experimental problem. Two beta sources (ionizing, radiation) are distributed in a sample of a certain thickness. The sources are confined to a certain volume. They radiate isotropically, but only a certain fraction of the radiation gets to the surface, which can be approximated as

$$I_1 \sim \frac{2\pi}{4\pi} \cdot \frac{\langle R \rangle}{D}$$

Where $\langle R \rangle$ is the mean range of the ionizing radiation and D is the thickness of the sample (see Table 13.1).

This radiation can be detected with the Berthold linear analyzer and digital autoradiograph. To be able to distinguish radioactive spots close to one another the detectors were made very thin (a few millimeters thick). They are position-sensitive multiwire chambers detecting with very high sensitivity and high spatial resolution. Compounds labeled with ^{3}H, ^{125}I, ^{14}C, and ^{32}P can be analyzed. Gamma-ray emitters, that is, ^{99m}Tc and ^{123}I are detected by gamma rays being converted into electrons by photo and Compton effect inside the counters.

LINEAR ANALYZER

The linear analyzer is well known (1–10). It can measure one- and two-dimensional radiation distributions (Figure 13.2). Beta particles and gamma rays emitted from the chromatogram enter the counter from the bottom

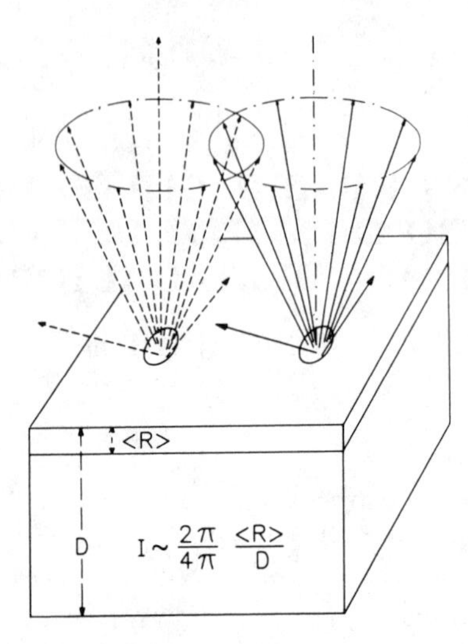

Figure 13.1. Beta emission of two separated compounds from sample of thickness *D*.

Table 13.1. Report of Measured and Analyzed Data

Plate ID...........:PLATE 1
Name of Track......:TRACK 1
Date of measurement:25-MAR-86 11:42:42

Runtime............: 0(hours) 1(min) 0(sec)
Position= 10.0(cm) RF-Start= 1.00(cm) Front= 20.00(cm) Gain=2
Significance= 3.0 Peak-reject= 30.(cts) Half-width= .3/ .0(cm)

# Name	RF	center (cm)	from (cm)	to (cm)	net (cts)	net (cpm)	S.D. (%)	ROIs (%)	all (%)
1	.16	4.09	3.74	4.45	822.8	822.8	4.08	2.32	1.89
2	.21	5.04	4.59	5.48	1824.6	1824.6	3.10	5.14	4.19
3	.24	5.61	5.23	5.98	1284.4	1284.4	3.93	3.62	2.95
4	.29	6.44	6.06	6.83	1135.3	1135.3	4.20	3.20	2.61
⋮									
14	.64	13.18	12.86	13.49	2159.8	2159.8	2.49	6.09	4.96
15	.70	14.31	13.93	14.68	786.5	786.5	3.95	2.22	1.81
16	.78	15.74	15.35	16.12	4179.9	4179.9	1.83	11.79	9.60
17	.80	16.23	16.00	16.46	459.2	459.2	5.82	1.29	1.06
18	.85	17.22	16.86	17.58	3527.4	3527.4	1.83	9.95	8.10

		(cts)	(cpm)
Total gross	:	43522.	43522.
Gross in ROIs	:	41407.	41407.
Net in ROIs	:	35465.	35465.

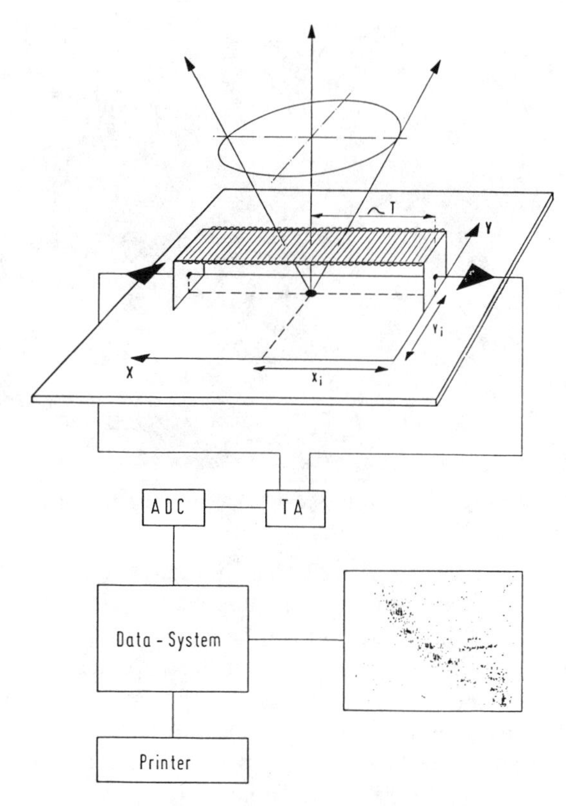

Figure 13.2. Principle of the linear analyzer.

through an open entrance window of 250-mm length and variable width, 1 to 25 mm. The position x_i, y_i of an ionizing particle emitted from the object plane (i.e., TLC plate) entering the detector is determined by measuring the projected position y_i of the counting wire and by measuring the time of propagation T of the pulse signal generated at the counting wire and transmitted into the delay line, and is proportional to the position x_i (see Figure 13.2). After appropriate electronic processing the signals are transmitted to the data acquisition system, an IBM-AT computer or equivalent. Figure 13.3 shows a photograph of the complete system: detector head, electronics for signal processing, and data acquisition system.

Linear or one-dimensional distributions, that is, radiochromatograms developed in one direction, are detected at once by placing the chamber directly above the chromatogram.

Following data acquisition, a peak search is performed. Peak fitting is done by a Gaussian fit method. This permits unfolding of overlapping peaks

Figure 13.3. Linear analyzer LB 285 with IBM-AT computer.

and results in an accurate representation of peak area, the background being automatically subtracted. Many overlapping peaks that cannot be resolved by the standard "drop line" method can be identified and quantitated. The results of the peak fitting and unfolding are then printed out in Gaussian curve format. This represents a dramatically improved method of evaluating and viewing the raw data. Bar graphs which translate the area of each peak into a relative height distribution are also produced. An example of this data analysis is shown in Figure 13.4 and Table 13.1.

Two-dimensional chromatograms are detected by moving the detector head in *Y* direction and by measuring in sequence and small steps down to 1 mm and up to several hundred lanes of the TLC plate or gel. From the acquired data the data acquisition system can generate on the display or on hard copy a two-dimensional position distribution of scale 1:1. This distribution, the digital autoradiography, is quantitative with a dynamic range of 1:32,000 per 0.025 mm^2, compared to the very limited dynamic range 1:10 of the classical autoradiography.

The computer analysis of the data goes far beyond the limitations of conventional autoradiography. Some aspects are described below. The spatial resolution of the detecting system is very high. In two- and three-dimensional distributions very closely positioned spots or overlapping spots can be recognized much easier than in linear distributions. Separations down to 1 mm can be detected. The detection efficiency of the system is remarkably high. Typically 1000 dpm of ^{14}C from a TLC plate (250 μm thick) can be detected in 1 min per lane, and a TLC plate of 20 × 20 cm with spot activities of 1000 dpm ^{14}C or more can be measured completely within 60 min.

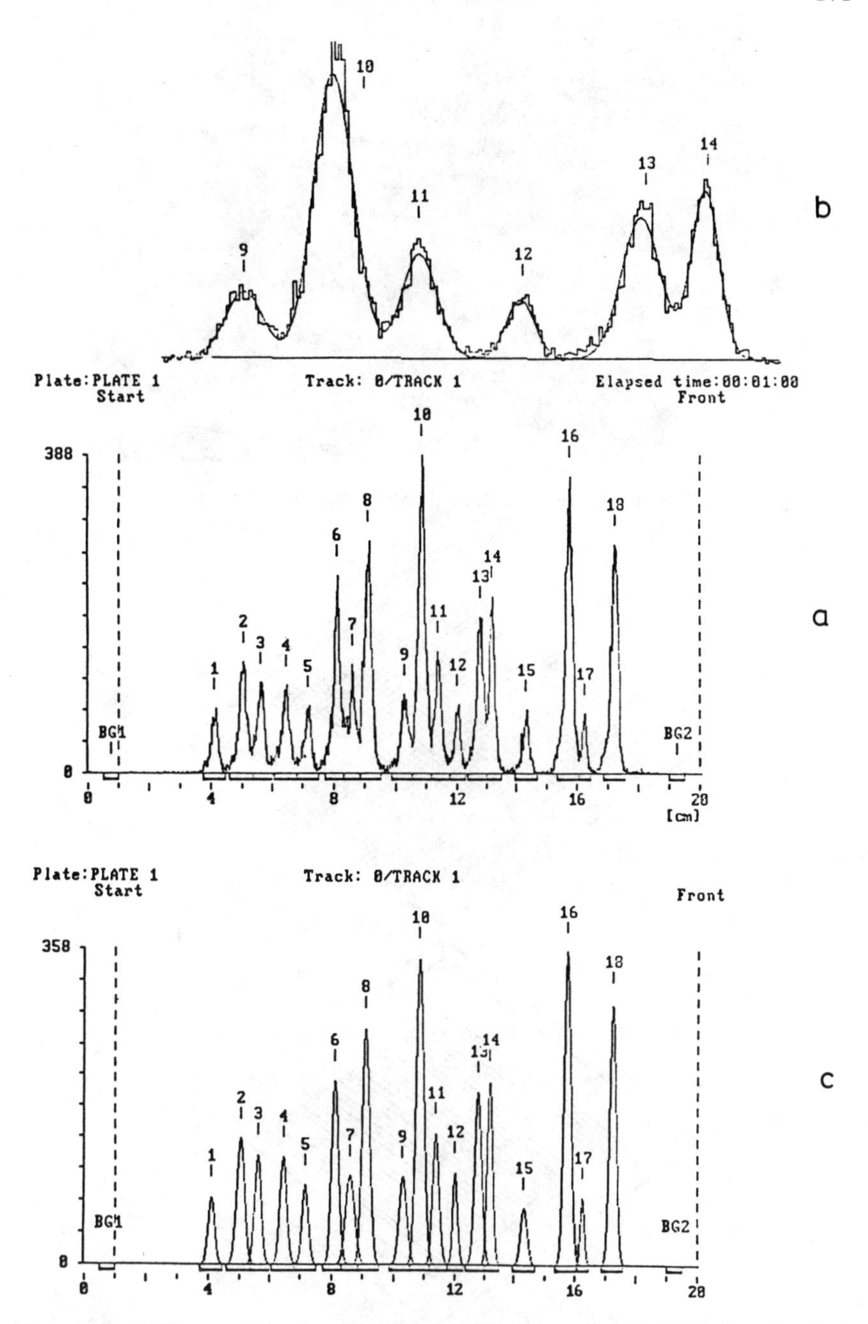

Figure 13.4. (*a*) Measured data from TLC plate. (*b*)Fit of data, Peaks 9–14. (*c*) Gaussian fit of total chromatogram.

Figure 13.5. Linear analyzer LB 284, modified for microtiter plate measurement.

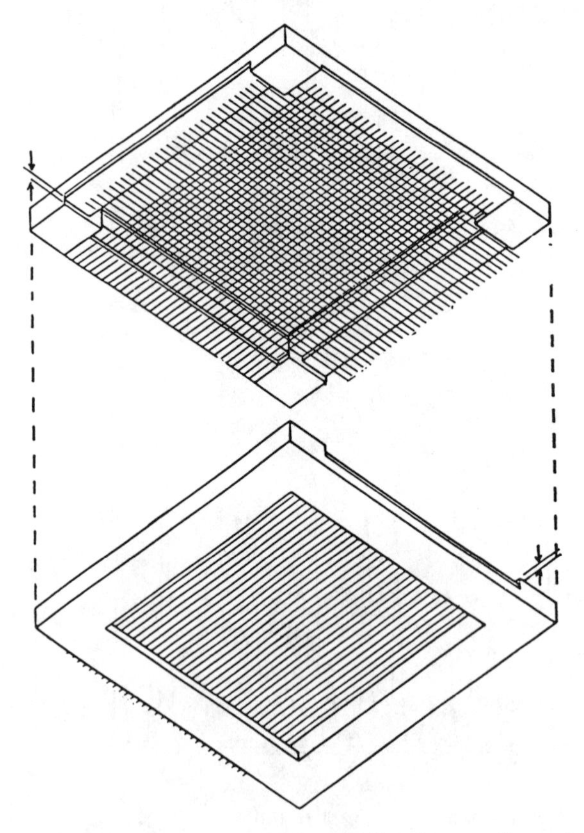

Figure 13.6. Schemataic view of wire planes of digital auto radiograph (DAR).

A new application of the linear analyzer is the measurement of the radiation from microtiter plates and filter mate. One plate contains 96 samples of about 5 mm diameter arranged in a matrix of 8 × 12. Instead of measuring each sample individually in a liquid counter, 24 samples are measured at once with the linear analyzer. With the LB 284 system 20 microtiter plates can be positioned on the measuring table and fully automatic measurement carried out, as shown in Figure 13.5. Special versions of the linear analyzer are available to measure the radiation of receptor assays, cell cultures, and DNA hybridizations bound on filter mate in combination with harvester systems, the applied isotopes being ^{125}I, ^{3}H, ^{35}S, ^{14}C, ^{32}P, and so on.

DIGITAL AUTORADIOGRAPH

This detector, recently developed, is a real two-dimensional, position-sensitive, multiwire proportional chamber (MWPC) with a 20 × 20-cm sensitive area. Very high sensitivity and high spatial resolution for compounds labeled with ^{3}H, ^{125}I, ^{14}C, ^{32}P, ^{99m}Tc, and so on, is possible. The detector consists of three wire planes, *X*, *Y* and *Z*, each with 100 wires (Figures 13·6 and 13.7).

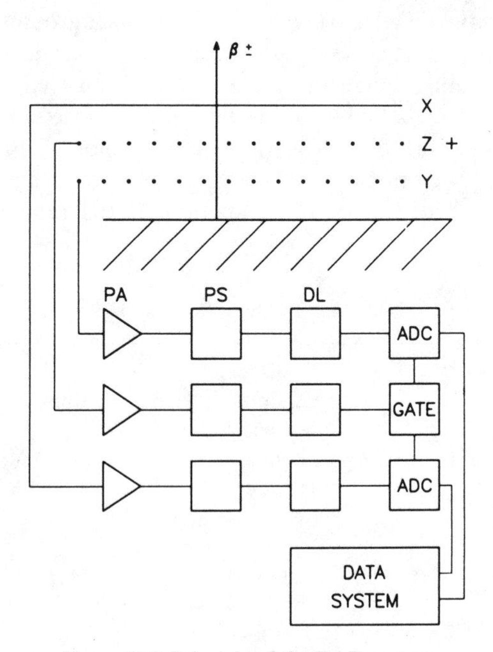

Figure 13.7 Principle of the DAR system.

The spacing between the planes and the wires is only 1 to 2 mm. The *Z* plane (middle plane) is at positive high-voltage potential, about 1700 v. The chamber is filled with P-10 gas (90% argon + 10% methane). The *Z* wires actually generate a charge signal the ionizing radiation entering the chamber. The two other orthogonal crossed-wire planes, *X* and *Y* below and above *Z*, pick up the charge signal from the *Z* plane and therefore determine the *X*, *Y* position of the traversing radiation, that is, a beta particle, as shown in Figure 13.7. The signals from the three wire planes, 300 wires, are transmitted via preamplifiers (PA), pulse shapers (PS), discriminators, and logic circuits (DL) to analog-to-digital converters (ADC) and finally to a powerful data-acquisition system, such as the Compac 386/20 computer, with color graphics.

Figures 13.8*a,b* show photographs of the complete system. The samples, TLC plates, gels, or tissue sections, are positioned on a measuring table, which automatically exposes them directly to the MWPC. The entire sample is measured at once. The acquired data are displayed on the CRT monitor of the data system as a *two-dimensional color distribution* of the emitted radiation. Digital autoradiography is quantitative with a dynamic range of 1:32,000 per 0.025 mm^2 as compared to the limited dynamic range of 1:10 of the classical autoradiography.

The two-dimensional analysis includes contrast enhancement and background subtraction to selectively suppress the background activity on the plate, which leads to higher resolution of the individual spots on the plate. By selecting the dynamic range for the evaluation, one can detect areas of activity normally overlooked in autoradiography, due to the limited dynamic range of photographic film. The radioactivity plot of the plate may be presented so that the optical density (degree of blackening, resp. color) is a linear, logarithmic, or square-root function of the radioactivity. A two-

Table 13.2 Technical Data of DAR.

Sensitive area	20 × 20 cm
Sensitivity	100 dpm ^{14}C and 1.000 dpm ^{3}H spot activities detected in 10 min
Position resolution	1 mm
Background	2 cpm/cm^2
Counting gas	90% argon + 10% methane gas flow 0.01 L/min
Dimensions	Height: 24 cm Width: 44 cm Depth: 44 cm

Figure 13.8. General view of DAR with data system.

dimensional peak search program allows fast peak recognition and integration of the radioactive labeled regions.

A three-dimensional radiation distribution of the sample can be displayed, that is, in the *X-Y* plane the position of the radiating spot and in the *Z* direction the amount of radiation emitted from this spot in color scale. The *X-Y* plane can be "viewed" under any angle and from any distance. The detection efficiency of the system is very high. Typically a TLC plate (250 μm thick) of 20×20 cm with spot activities of 1000 dpm ^{14}C or more can be measured completely within 1 min.

Table 13.2 summarizes some technical data. The *Figures 13.9–13.15* show some examples of measured and analyzed data using the digital autoradiograph.

APPLICATION

The new digital autoradiograph has the same or similar applications as the linear analyzer, except it is much more sensitive and faster. A two-dimen-

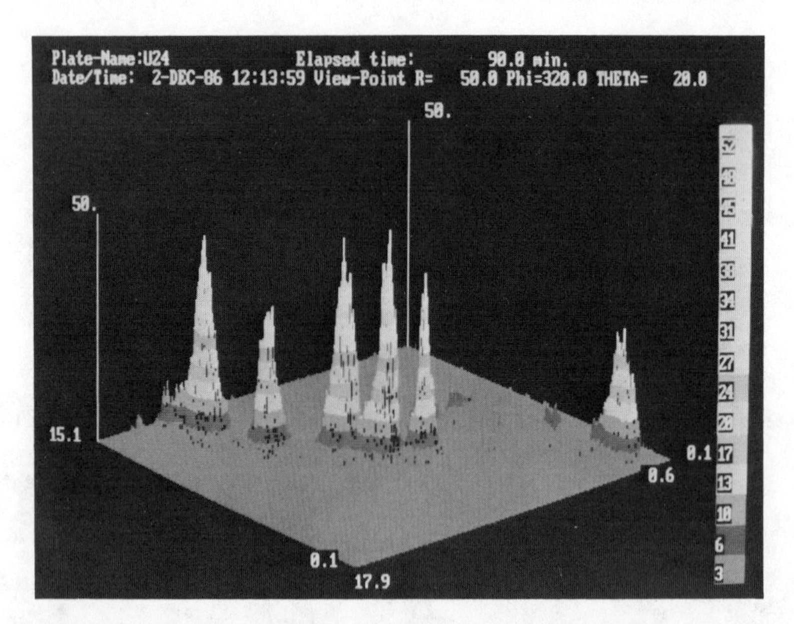

Figure 13.9. (*a*) Display of two-dimensional chromatogram with DAR. The color scale on the right is proportional to the measured activity of ^{14}C. Total ^{14}C activity on the TLC plate is about 5000 dpm measured in 30 min. (*b*) Three-dimensional presentation of (*a*), intensity against position.

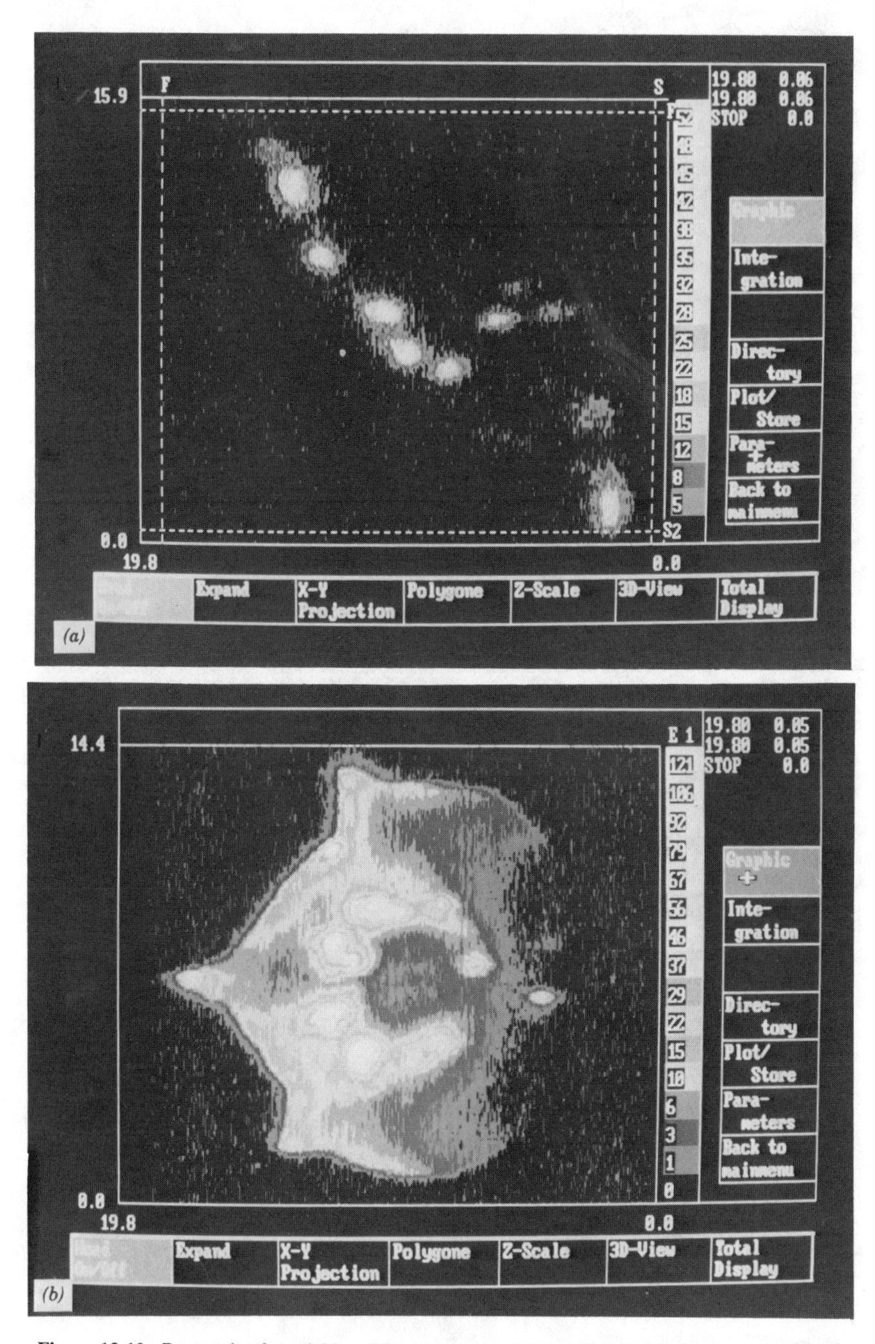

Figure 13.10. Research of pesticides. Measurement of ^{14}C radioactivity from a cotton leaf.

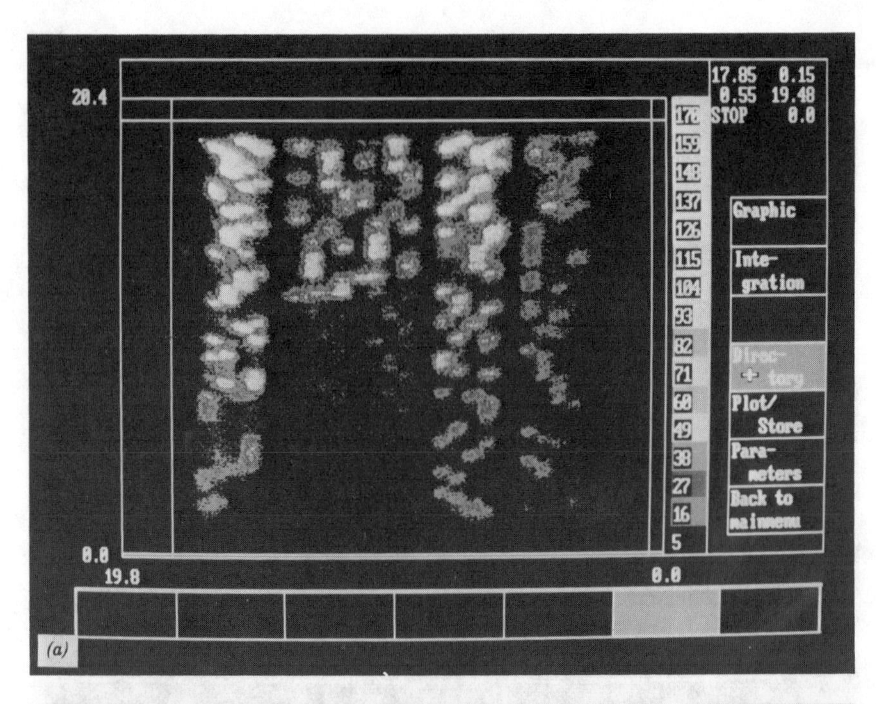

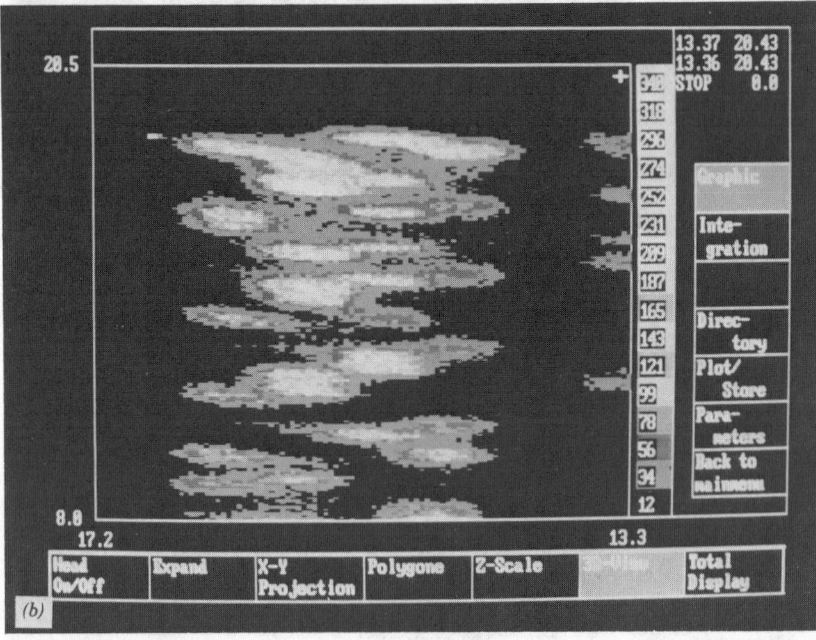

Figure 13.11. Measurement of DNA sequences labeled with ^{35}S. (*a*) Complete gel; (*b*) enlarged section of upper left part.

Total measuring time (min.):	616.4
Gain:	1
Backbias:	.00000
Z-factor:	1.3000
Z-scale:	Lin

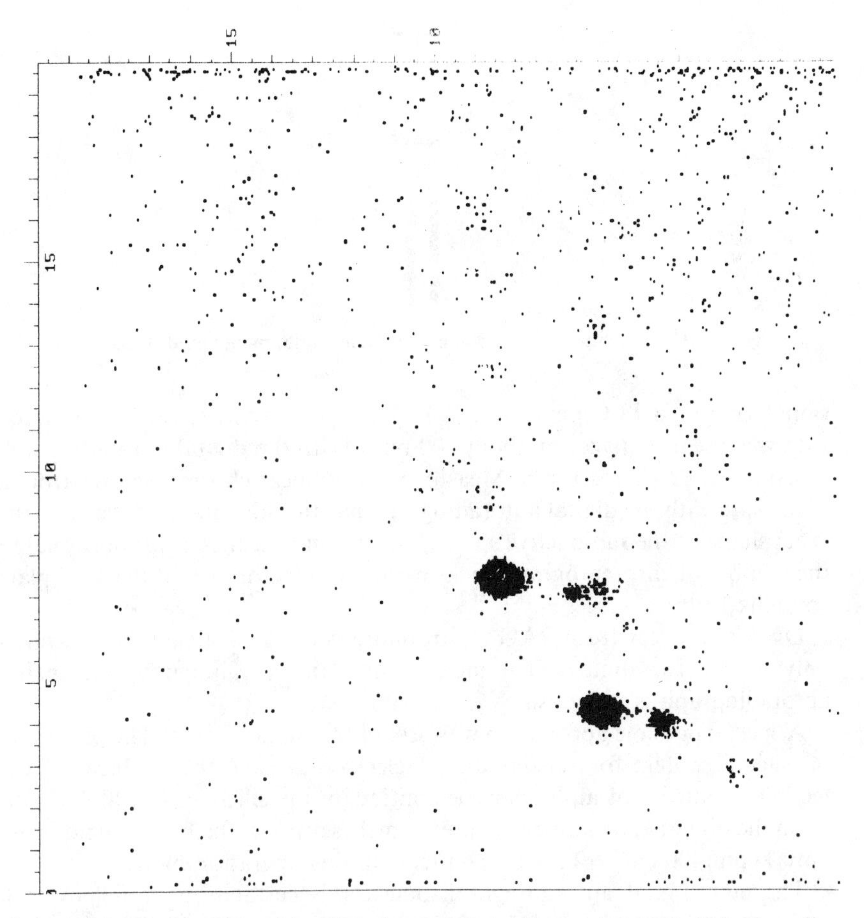

Figure 13.12. Very low ^{14}C activity TLC plate (0.5–30 cpm), measured within 10 h. X-ray film exposure took 4 months to detect the low activity.

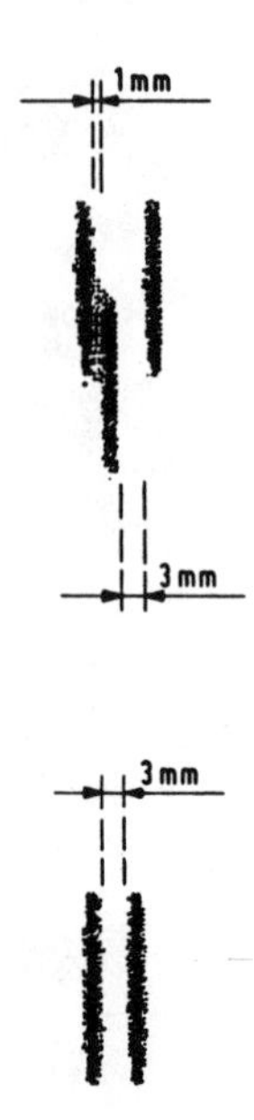

Figure 13.13. Resolution of ^{14}C sources of 1 mm width, minimum distance 1 mm.

sional sample, a TLC plate with 1000 dpm spot activities, can be measured with the linear analyzer in about 100 min. With the digital autoradiograph measurement requires 1 min. Measurement of linear chromatograms from a TLC plate with the digital autoradiograph has the advantage of "seeing" the exact shape of the radioactive spots. The gain in time here is proportional to the number of chromatogram tracks per plate, a factor of 10 if the TLC plate contains 10 tracks.

DNA sequences from PAG electrophoresis can realistically be measured only in the two-dimensional mode. Thus, the advantage of the digital autoradiograph is obvious.

A special possible application was brought to our attention. The producers of silicone wafers for semiconductor electronics have the problem of detecting impurities of alpha-particle emitters in the silicone. For alpha radiation the digital autoradiograph is extremely sensitive, the background being only 1 cpm/400 $cm^2 = 2.5 \times 10^3$ cpm/cm^2 in this operation mode.

The new digital autoradiograph detector is certainly a more powerful "microscope", which should open new areas as research is progressing. Typical fields of application are:

1. Pathways of metabolites (TLC)
2. Purity control of radiopharmaceuticals (TLC)

Total measuring time (min.):	922.0
Gain:	4
Backbias:	.00000
Z-factor:	1.0000
Z-scale:	Lin

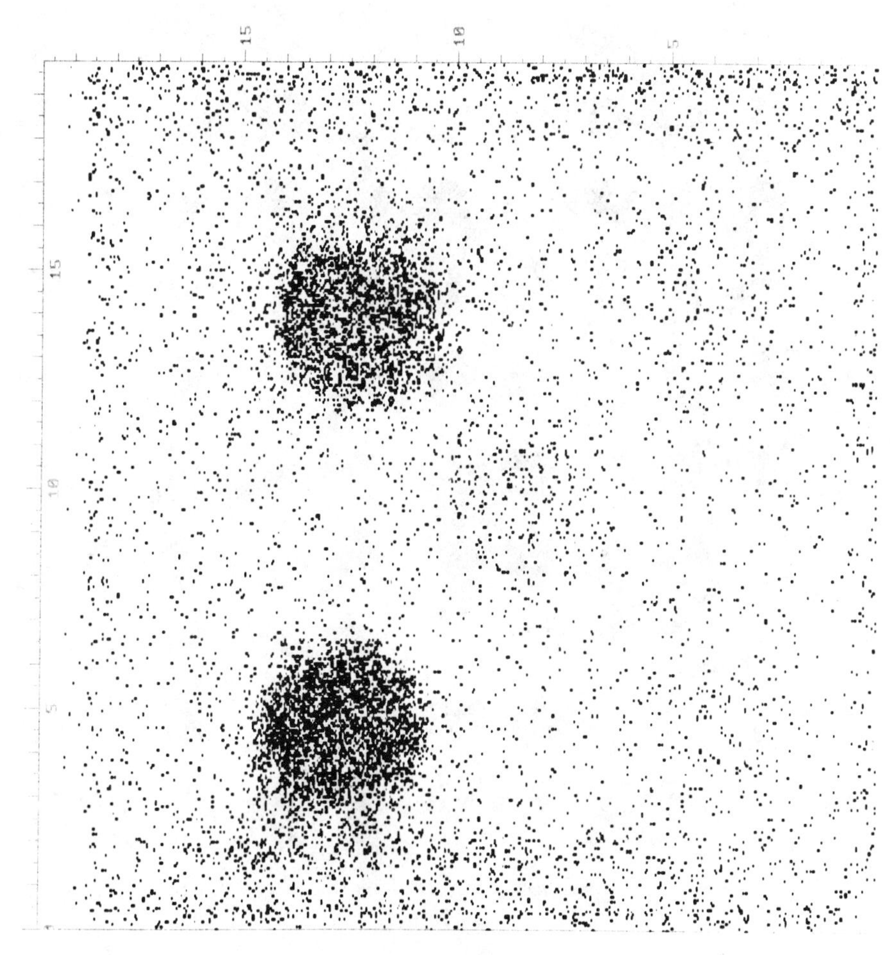

Figure 13.14. Measurement of alpha particles from semiconductor samples. The active samples have about 0.03 cpm/cm^2, the background is 0.0025×10^2 cpm/cm^2.

```
Total measuring time (min.):        5.4
Gain:                             1
Backbias:                           1.0000
Z-factor:                           1.0000
Z-scale:                          Lin
```

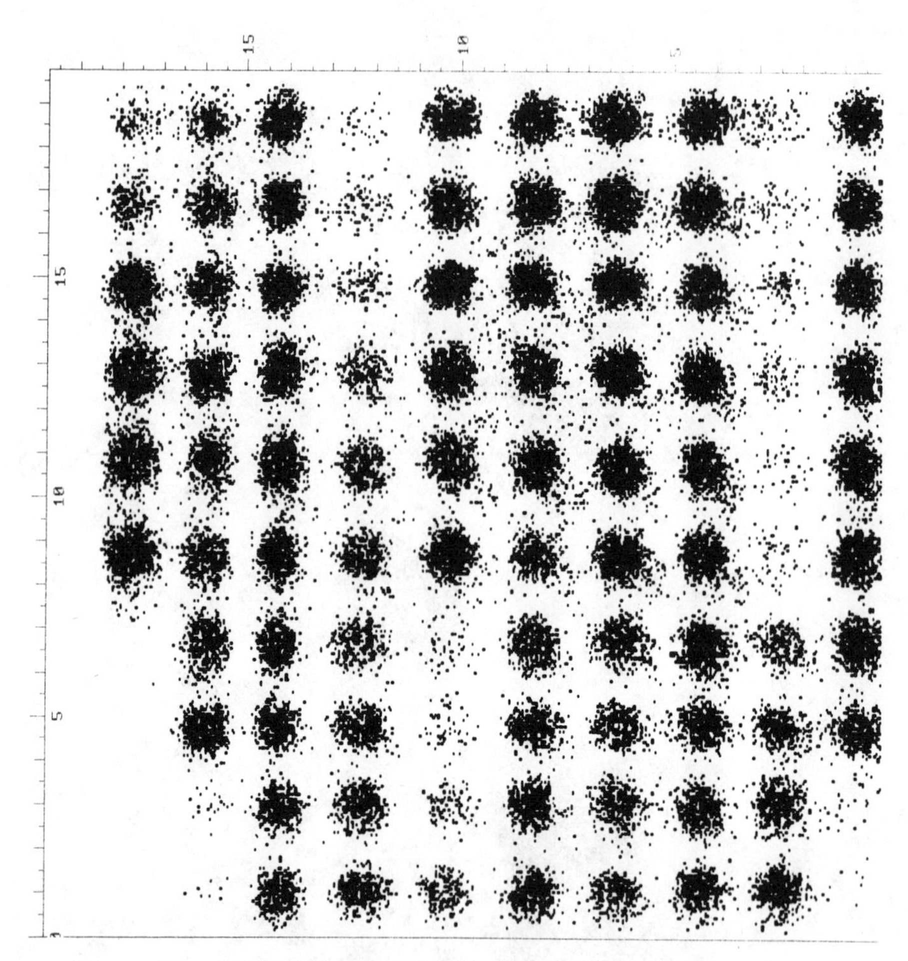

Figure 13.15. Measurement of receptor binding assay, ^{14}C label.

3. DNA probes, DNA hybridization
4. Receptor binding assays, cell cultures on filter mate (harvester samples)
5. Microtiter plates
6. DNA sequences, PAG electrophoresis

7. Protein identification, blotting
8. Tissue sections
9. Control of material (thin layers, film) against radioactive impurities, such as alpha emitters in semi conductors

ACKNOWLEDGMENT

I thank W. Schuh very much for his support in developing the data analysis software.

REFERENCES

1. H. Filthuth, *Indust. Res. Dev.* (June 1981).
2. H. Filthuth, *Synthesis and Applications of Isotopically Labeled Compounds*, Proceedings of an International Symposium, Kansas City, MO, 6–11 June, 1982, W.P. Duncan and A.B. Susan (Eds.), Elsevier, Amsterdam, 1983, pp. 447–452.
3. H. Filthuth, *Advances in Thin Layer Chromatography*, J. C. Touchstone (Ed.), Wiley, New York, 1982, Chap. 7.
4. H. Filthuth, *Techniques and Applications of Thin Layer Chromatography*, J. C. Touchstone (Ed.), Wiley, New York, 1985, Chap. 10.
5. H. Filthuth, *Analytical and Chromatographic Techniques in Radiopharmaceutical Chemistry*, D.H.Wieland, T.J. Mangner, and M. Tobes (Eds.), Springer-Verlag, New York, 1985, Chap. 4.
6. E.A.M. Fleer, D. -J. Kim, C. Unger, and H. Eibl, Metabolism of 1-0-Octadecyl-2-methoxy-rac-glycero-3-phosphocholine in Raji cells, Max Planck Institut für Biophysikalische Chemiev, Göttingen, preprint 1985.
7. A. Hammermaier, E. Reich, and W. Bögl, IHS-Report 36, Institut für Strahlenhygiene des Bundesgesundheitsamtes, Neuherberg/München, February 1984.
8. A. Hammermaier, E. Reich, and W. Bögl, IHS-Report 48, Institut für Strahlenhygiene des Bundesgesundheitsamtes, Neuherberg/München, August 1984.
9. P. Jamet and J. -Ch. Thoisy, Evaluation et Comparaison de la Mobilité de différents Pesticides par Chromatographie sur Couche Mince de Sol, Service de Chimie des Pesticides, Département de Phytopharmacie et d'Ecotoxiologie I.N.R.A., Versailles, preprint 1985.
10. H. Filthuth, *Synthesis and Applications of Isotopically Labeled Compounds 1985*. Proceedings of the Second International Symposium, Kansas City, MO, 3–6 September 1985, R.R. Muccino (Ed.), Elsvier Science, Amsterdam, 1986, pp. 465–472.

CHAPTER

14

NEW TECHNIQUES IN TWO-DIMENSIONAL DATA PROCESSING

DEAN E. SEQUERA

This paper reports the results of a study of the measurements of two-dimensional electrophoretic and thin-layer chromatographic (TLC) patterns. Two qualitative mapping techniques are shown. Quantitative analysis of partial spots using higher order calibration curves is also examined.

Two-dimensional planar chromatography has been a growing field of analytical chemistry. This technique is capable of producing simultaneous separation and identification of a large variety of samples. The techniques include high-performance thin-layer chromatography (HPTLC), electrophoresis, and paper chromatography. The resolving power is far superior to rival techniques such as TLC. But, when compared to HPLC, the reproducibility has been relatively poor, quantitation has been limited, and automation is difficult. Qualitative visualization of complex patterns is important for interpretation. But the necessary hardware and software expense to achieve interpretable results eludes the budget of most research laboratories.

A tremendous sample load has been placed on many analytical laboratories for drug screening. HPTLC offers very high throughput with the ability to apply many samples on one plate. In the past, a specific method for each drug of interest was developed due to the limitations of the detection technique. Thus, many sample preparations were necessary, which somewhat offset the advantage of multiple spotting. Accurate quantitative analysis has been difficult as well due to sample detection and data processing limitations.

Video scanners with a television-type camera combined with image analysis software have achieved great speed of analysis, but have some inherent disadvantages in the low-resolution, narrow dynamic range, and poor signal-to-noise ratios. On the other hand, single spot beam scanners using photomultipliers have a wide optical range, excellent resolution, but very slow scanning speeds because of point-by-point measurements.

NEW INSTRUMENTATION

The Shimadzu CS-9000 dual wavelength flying spot scanning densitometer was used for these analyses. The CS-9000 consists of a spectrophotometer with an *X-Y* plate mounting stage, a controller/data processor with a high-resolution thermal plotter, and a video display.

The spectrophotometer is a Monk-Gillieson type with a 600-groove/mm grating. There are 16 possible exit slits which determine the dimensions of the beam at the sample plate mounted on a rotary disk. The measuring wavelength ranges from 200 to 700 nm. Two light sources are available, a deuterium lamp for the UV region and a tungsten–halogen lamp for the visible region. Optional sources include a helium–neon laser, xenon lamp, or mercury lamp, the latter two finding use primarily in fluorescent techniques.

The system may measure transmittance, reflection absorbance, or fluorescence using the photomultiplier above the sample or directly below. A beam splitter is employed before the beam reaches the sample and directed toward a reference photomultiplier. The reference and sample photomultiplier outputs are ratioed to produce the measurement signal. A log converter is used to produce an absorbance signal.

The measuring mode may be a strict linear scanning of the sample stage as is found in conventional densitometers. In addition, the system may use the high-performance zigzag, flying spot method in which the fixed and rotary

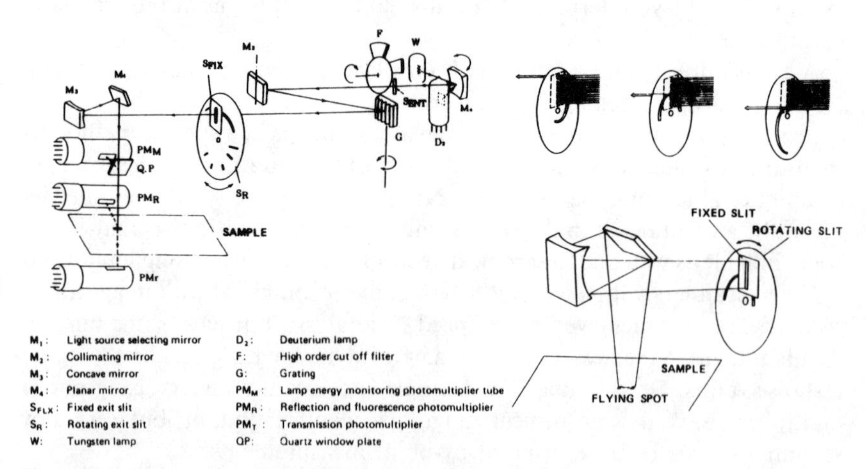

Figure 14.1. Flying spot principle. The light beam is masked by the curvelinear segment in the rotating slit S_r and in combination with the fixed slit S_f the beam is deflected along the *X* axis. As the sample stage moves in the *Y* direction, a zigzag pattern is developed.

slits are combined. This technique is similar to techniques found in other analytical instruments such as scanning electron microscopes. The major advantage to this technique is that the entire area of a spot may be integrated.

By rotating the rotary slit relative to the fixed slit, an optical flux is produced at the sample measuring site as shown in Figure 14.1. The sample stage movements are coordinated with one complete cycle of the moving slit by a stepper motor in the *Y* direction.

QUALITATIVE AND QUANTITATIVE EXAMINATION OF DIFFICULT SAMPLES

Two-Dimensional Gel Electrophoresis

Two-dimensional gel electrophoresis involves a great deal of sample handling and sources of error. The variables include gel thickness, uniformity, stain application, and destaining rinse. Four techniques of electrophoresis have found common use: moving boundary, isotachophoresis, zone, and isoelectric focusing. The two-dimensional nature involves combining two of these different separating principles.

For example, the most popular technique, first discussed by O'Farrell in the 1970s, subjects the sample to one-dimensional isoelectric point electrophoresis in a polyacrylamide tube gel. The technique is illustrated in Figure 14.2. A pH gradient is established across the gel and the proteins in the gel move to a point where the net charge at that pH is zero (isoelectric point). After the separation is complete, the gel is removed from the tube and placed in close contact with a sodium dodecyl sulfate (SDS) gradient slab gel. Then a zonal electrophoretic process is performed perpendicular to the polyacrylamide gel. The SDS gel denatures the proteins and acts as a molecular sieve, separating them by molecular weight. Thus, this two-dimensional process characterizes the proteins in the sample on the basis of both their isoelectric points and molecular weights. The gel can be stained at this point for characterization and quantitative analysis using a silver stain or Coomassie brilliant blue.

Figure 14.3 shows a two-dimensional electrophoretic gel that was used for this experiment. This silver-stained sample contains the total proteins in human blood platelets. The horizontal separation indicates the isoelectric point, while the vertical indicates the molecular weight. The transmission spectra of the silver-stained spots were obtained using the wavelength scanning function of the spectrophotometer. It was found that the optimal wave length with the maximum absorbance and sensitivity was 420 nm. The tungsten–halogen source was used for maximum energy at this wavelength.

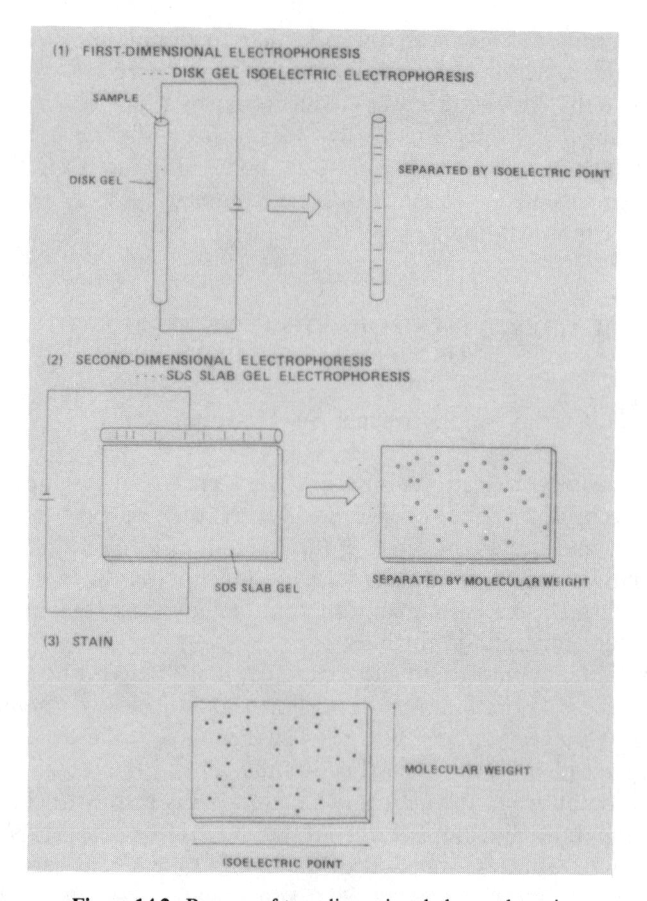

Figure 14.2. Process of two-dimensional electrophoresis.

The optimal wavelength for Coomassie brilliant blue stain was found to be 500 nm.

The CS-9000 possesses three mapping modes to record the two-dimensional pattern of the gels and TLC plate. The contour mode produces a mark where the absorbance exceeds an operator-set threshold value. The hatch mode blots out all measurements above this threshold value. The contour map produces a mark at the threshold value and at operator-set increments above this value.

Figures 14.4 and 14.5 show examples of this mapping function using the gel shown in Figure 14.3. Each mapping was accomplished in 20 min. The maps may now be archived for future reference and consulted for X and Y coordinates for specific spots. Ten different protein spots were selected for

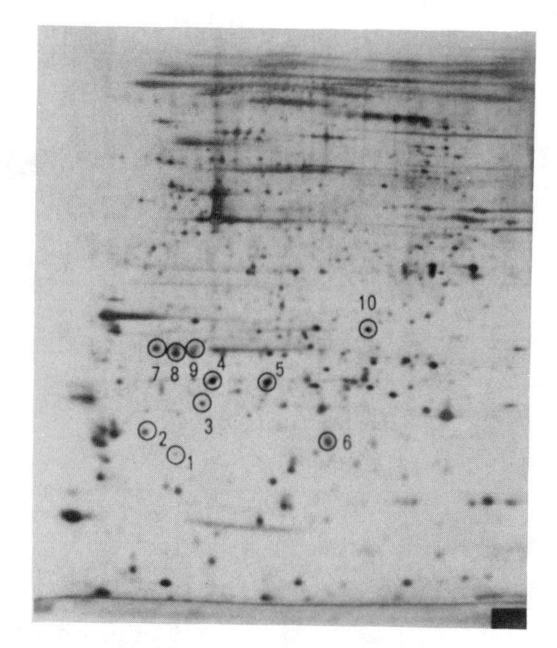

Figure 14.3. Measured spot.

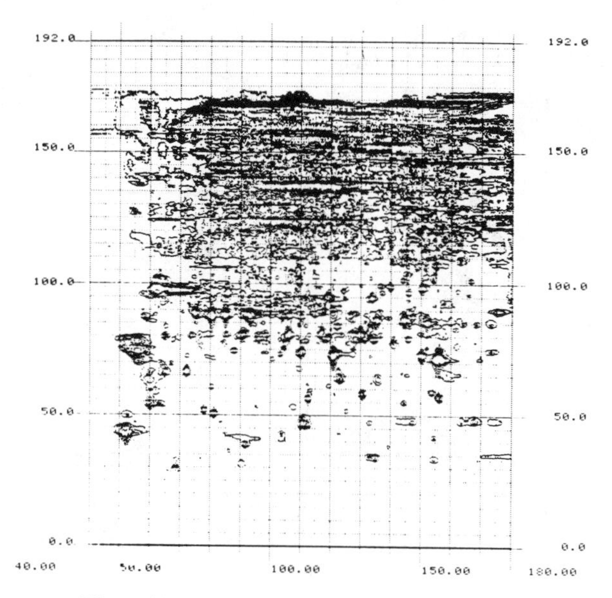

Figure 14.4. Gel mapping by contour map mode.

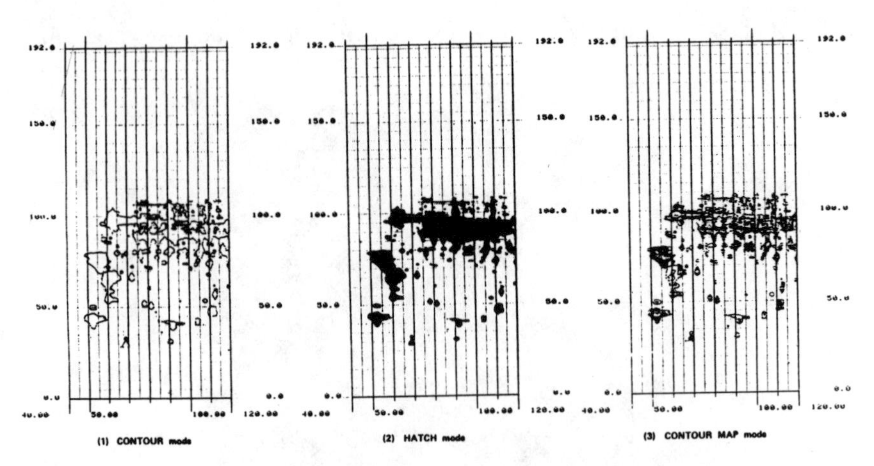

Figure 14.5. Mapping record.

NO.	SPOT CHROMATOGRAM	NO.	SPOT CHROMATOGRAM
1	0.600 0.600 0.000 0.000	6	0.600 0.600 0.000 0.000
2	0.600 0.600 0.000 0.000	7	0.600 0.600 0.000 0.000
3	0.600 0.600 0.000 .000	8	0.600 0.600 0.000 0.000
4	0.600 0.600 0.000 0.000	9	0.600 0.600 0.000 0.000
5	0.600 0.600 0.000 0.000	10	0.600 0.600 0.000 0.000

$\overline{X}$ = 32513.67

S.D. = 197.20

C.V. = 0.61 %

Figure 14.6. Reproducibility of spot area value.

study across the four gels. These 10 proteins are well studied in blood platelet research and reflect a wide range of dimensions, concentrations, and separation from other spots. The instrument allows control of the acquisition signal, automatic or manual area calculation, and other data processing options.

The reproducibility of the measured values of a single spot was first studied. A spot was measured 10 times. The mean area was found to be 32,513.67 counts. The standard deviation was 197.20 and the coefficient of variation was 0.61%, which indicates excellent reproducibility (see Figure 14.6). All 10 spots across the four gels were then measured. When one considers an individual gel as a whole, there may be differences between the gels because of variables that affect the entire gel. This was overcome by normalizing the area counts such that the average value of the 10 spots was 100. The results are summarized in Table 1.

Using these values, the correlation coefficients of the results between gels were calculated. Plots of the six combinations are shown in Figure 14.7. All of the calculated correlation coefficients exceeded 0.95, indicating excellent correlation among them. One could see with visual inspection that the same spot between gels showed marked differences in shape and dimensions due to the media and preparation. The fact that excellent correlations were achieved

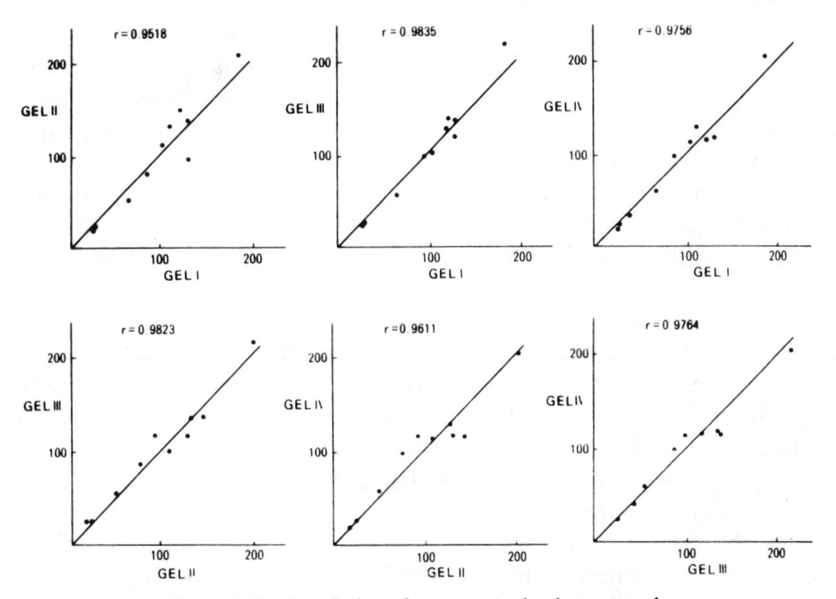

Figure 14.7. Correlation of spot area value between gels.

Table 14.1 Evaluation of Reproducibility

	Gel I	Gel II	Gel III	Gel IV
No. 1	28.7	23.3	26.2	24.1
No. 2	87.4	78.0	87.6	98.2
No. 3	27.1	19.2	24.4	19.9
No. 4	105.3	110.5	100.6	113.3
No. 5	111.8	130.8	117.7	129.7
No. 6	132.2	95.1	118.0	117.7
No. 7	122.9	148.9	138.9	115.2
No. 8	187.3	207.4	218.4	205.3
No. 9	131.2	135.4	136.3	117.5
No. 10	66.1	51.4	55.4	59.2

nonetheless demonstrates the flying spot method's ability to compensate for this irregularity.

HPTLC

A complex multilane HPTLC plate was also examined to establish the feasibility of using this new technique for drug screening of urine samples. This plate had 30 lanes of 20 various drugs, including quinine, nicotine, cocaine, chlorpheniramine, dextromethorphan, benadryl, brompheniramine, doxepin, methyldoxepine, imipramine, desipramine, amoxapine, amitriptyline, noramitriptyline, ephedrine, pseudoephedrine, and amoxapine. Various volumes were spotted. The plate was developed with a 90:30:1 mixture of ethyl acetate:cyclohexane:isopropylamine. All compounds and metabolites are known to absorb in the UV region for detection, although the absorbance maxima vary.

No single wavelength was found to be universally acceptable but analysis at 240 and 260 nm were found to adequate to detect all components using the reflectance absorbance mode. Figures 14.8 and 14.9 show a hatch mapping of the two scans. Each mapping required 18 min. At a quick glance, the operator can determine a positive identification. If quantitation is desired, standards can be spotted in the first lanes and a calibration curve can be automatically developed.

The CS-9000 may create a calibration curve from automatically calculated area counts. The operator may choose linear, semilog, log–log, linear-log logit, or log–log logit axes. A first-, second-, or third-order fit may be

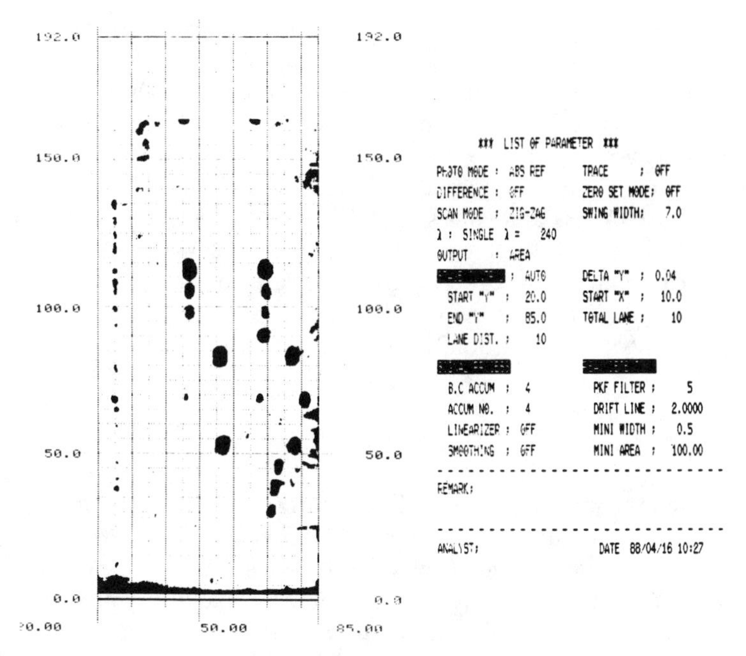

Figure 14.8. Hatch mapping at 240 nm.

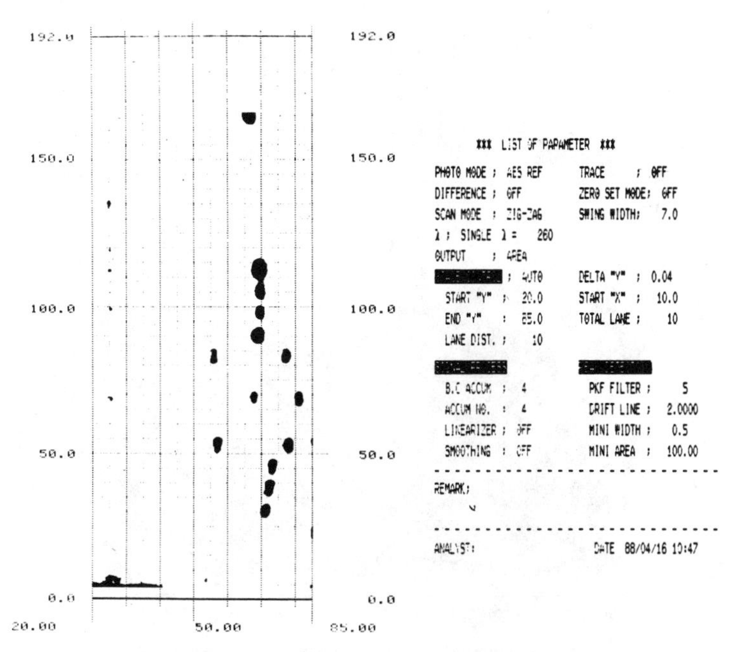

Figure 14.9. Hatch mapping at 260 nm.

determined with up to 10 standards. Several calibration curves may be linked should the analytical wavelengths and extinction coefficients vary.

Two-dimensional analysis with HPTLC plates is much easier and much more accurate compared to electrophoretic gels. The medium is extremely uniform in thickness and consistency by comparsion. Quantitative analysis with this medium is well characterized, with the major deviation from linearity caused by light scattering from the surface which is not collected at the photomultiplier. With the CS-9000, this can be automatically corrected using the Kubelka-Munk family of transformation curves.

DATA PORTABILITY

In the modern information-handling environment, there is an increasing need to archive, analyze, and report data using a microcomputer. Modern digital instrumentation allows the data to be transferred to a computer for special requirements. The state of software available for microcomputers today makes this task rather straightforward and powerful.

Figure 14.10 shows a two-dimensional rendition of a portion of the TLC plate analyzed above. The hardware used was the IEEE-488 GPIB parallel

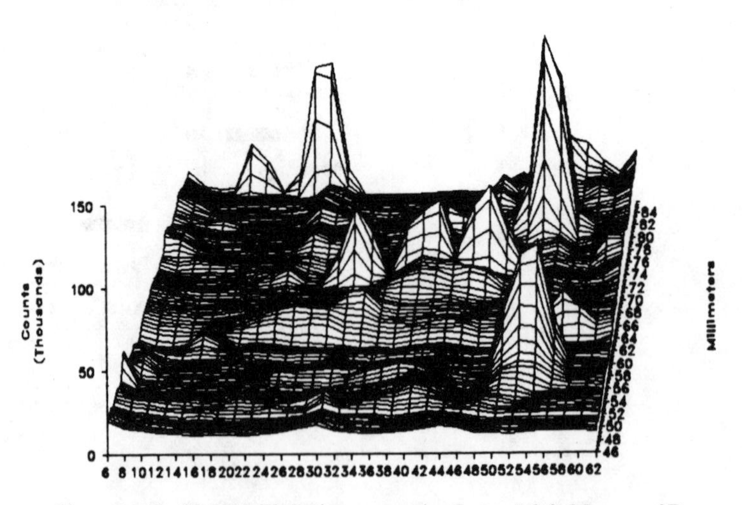

Figure 14.10. CS-9000 TLC plate map using Lotus 1-2-3, Measure, 3D.

interface of the CS-9000 and an IBM PC-AT with a National Instrument Co. PC-488 IEEE-488 interface. The data was collected using Lotus Measure and Lotus 1-2-3 from Lotus Development Corp. The three-dimensional software was an add-in product for Lotus 1-2-3 called 3D Graphics by Intex Solutions Inc. The data acquisition is extremely fast and easy. Commands are issued to the instrument via embedded ASCII strings within Measure's command protocol. Table 14.2 shows the Lotus 1-2-3 macros used to collect this data and insert a chromatogram scan into the worksheet. Once resident, the data can be graphed using 3D graphics as shown in Figure 14.10. On a color monitor and a color plotter, the ordinate data can be sectioned into color areas showing the ordinate level of peaks quite clearly.

Results of an important experiment can be communicated clearly with this output. The data can be stored permanently for later recall and analysis. Data can be manipulated within the worksheet. Two files may be combined and added together. For example, components may have very different extinction coefficients so a plate should be scanned at two different wave-

Table 14.2. CS-9000 TLC Plate Data Using Lotus 1-2-3, Macros (19 Apr 88, 07:35 a.m.)

	A	B	C	D	E
1	{CLEAR CS9000}		6	8	10
2	{OUTPUT CS9000, "START"}	20	21271	17304	14621
3		20.4	21284	17247	14983
4	{OUTPUT CS9000, "SEND D-FILE"}	20.8	21283	16262	15546
5	{NREAD CS9000, B1 . . B2000, 2000}	21.2	21085	16240	15593
6	/WIC~	21.6	20720	15713	15079
7		22	20484	15786	14221
8		22.4	20359	16017	13795
9		22.8	21173	16239	13350
10		23.2	22983	16674	13104
11		23.6	24262	17999	13278
12		24	26718	19651	13690
13		24.4	30042	21867	14432
14		24.8	32691	24578	15580
15		25.2	32561	26809	16037
16		25.6	29700	25501	14997
17		26	28431	23405	13951
18		26.4	28637	21904	13589
19		26.8	31916	21446	13850
20		27.2	34187	19971	14085

lengths. Then the two separate files may be combined, added, and the result displayed to show all components of interest.

SUMMARY

It was shown that the mapping function of the CS-9000 scanning densitometer allows the quick recording of complicated two-dimensional patterns. It was also confirmed that the reproducibility of the quantitative analysis of the spots was excellent even in a difficult electrophoretic gel with the CV of 0.6%. The correlation coefficient of quantitative values of spots among gels exceeded 0.95 by normalizing the total spot area value. Tedious drug screening analysis time can be shortened using the flying spot scanning technique. Modern microcomputers with popular software can be quite powerful in illuminating and reporting results. This software is very simple and inexpensive. The flying spot method should be a valuable technique for fast and precise measurements of two-dimensional planar chromatograms.

INDEX

Accelerated chromatography, 1
Acid stable fluorescence indication, 31
Adenine, 34, 35
Adenosine monophosphate, 34
Alanine oligomers, 12
Amersham test plate, 142
Ammonium biomide, 41
Androgens, 39
Anion exchange cellulose, 49
Antiganglioside antibody, 59
Array detector, 139
Artifacts, 69
Ascorbic acid, 119
  HPLC, 120
  radioscan, 123
Autoantibodies, 61
Autoradiograms, 61
Autoradiography, 128

Barbiturates, 24
Beta particles, 129, 167
Bile acids, 16
Biphenyl, 101
Blood, 120, 125
Butyric acid, 49, 53

C^{14}, 127
Calibration standards, lipids, 12
Centrifugal chromatography, 1
Chamber saturation, 72, 73
Chlorophenols, 30
Cholesterol, 13
Cholesteryl ester, 13
Cholic acid, 16
Cinchonine, 85
Collimators, 149, 150, 151, 152, 154,
Concentrating zone, 28, 29
Contour mapping mode, 188, 189
Cyclodextrin, 81, 82, 83
  α, 82
  β, 82
  hydroxypropyl, 86
  hydroxyethyl, 86
Cytochrome oxidase, 101

Dansyl amino acids, 81, 83, 88
Dansyl-DL-aspartic acid, 7
Dansyl-DL-glutamic, 95
Dansyl leucine, 95
Decomposition on plate, 75
Dehydroascorbic acid, 119
  blood, 120
  dinitrophenylhydrazone, 120
  seminal fluid, 120
Densitometry, 51, 103, 114, 121, 122, 186, 194
Detector array, 139
Diasteriomers, 89
Diethylstilbesterol, 17
Digital autoradiograph, 173
2-dimensional electrophoresis, 187
2-D process, 188
2D-scanning of isotopes, 4
Dinitrophenylhydrazone, 120

Electrophoresis, 187
Enantiomer separation, 81
17β-estradiol, 38
Estrogens, 38
17α-ethinyl estradiol, 38

Fats, 29
Feces, 102
Fermentation feedstock, 49
Ferrocene enantiomers, 90
Fluorescence, 2
Fluorescence quenching, 103
Flying spot densitometer, 185, 186
Fourier transform infra red spectrometry on TLC, 157

Gamma emitter, 148
Gamma rays, 169
Gangliosides, 64, 65
Geiger counters, 139
β-glucuronidase, 102
Glycolipids, 62, 63
Gramicidin, 36
Guanine, 34, 35
Guanosine monophosphate, 34

Hatch mapping mode, 192, 193
Hydrocortisone, 32
Hydrophobic, 82
Hydrophylic, 17, 82
4-Hydroxybenzoic acid, 2
2-Hydroxybiphenyl, 101, 104
4-Hydroxybiphenyl, 101, 104
Hydroxylation activity, 101

Immunoenzyme methods, 59, 60, 61
Immunostaining, 59
Induction, 101
Infrared, 157
Infrared spectrometry, 157, 162
In situ reaction, 112
Ion exchange, 2
Ion exchange cellulose, 49
Ion exchange SPE, 2
Ionizing radiations, 167
Ion pair, 42
Ion pair reagents, octane-1-sodium sulfonate, 42
Isoelectric point, 187
Isomers, structural, 89
Isonicotinic acid, 27

Linear analyzer, 169, 170
Lipid peroxidation, 111
Lipids, 8, 12
Lithium chloride, 25
Low energy gamma emitter, 149
Low temperature, 30
Luminal, 24
Lysine-PTH, 40

Malondialdehyde, 111
Mapping function, 188
Medroxyprogesterone, 45
Mesohexestrol, 45
Methyl prednisolone acetate, 162
Methyl testosterone, 32
Migration characteristics, 21
Migration time, 11, 21
Modern day geiger counter, 139

Neuraminidase, 64
Neuroblastoma cells, 66
Nicotinic acid, 27
^{99m}TC, 149
Nite red, 7, 8
Normal phase SPE, 2
19-nortestosterone, 45
Nucleosides, 35

Oils, 29
One dimensional scanning for isotopes, 128
Optitrain, 159, 160
Organic modifier, 82
Overpressured thin layer chromatography, 3

^{32}P, 130
Pentapeptides, 27
Peptides, 27
Phenobarbital, 101
Pheochromocytoma cells, 64
Phosphatidyl choline, 164
Phosphatidyl ethanolamine, 164
Phosphatidyl glycerol, 164
Phospholipids, 2, 164
Plasma, 101, 125
Polarity, 43
Polycyclic aromatic hydrocarbons, 31
Polyisobulytmethacrylate, 60
Polyrinyl pyrolidone, 59
Poly-X-antigens, 61
Position sensitive proportional counter, 138, 146
Preparative, 4
Progesterone, 45
Proportional counters, 4
Pulse height analysis, 130
Pure volume, 17
Pyrene, 86

Quantitative analysis, 187

Radio scanning, ^{32}P, 130
Removal, 24
Reproducibility, 192
Resolution *vs.* concentration, 90
RP-2, 20, 32
RP-8, 20, 32
RP-18, 20, 32

Retention, analyte, 2
Reversed phase, 15, 17
Reversed phase, SPE, 2

Sample application, 28
Selectivity, 32
Semen, 125
Seminal fluid, 120, 125
Separation efficiency, low temperature, 30
Separation efficiency increase, low temperature, 30
Shift of absorption maxima, 104, 105
Sialosyl-Le-antigen, 8
Silica gel, 51
Silica gel, water content, 74
Silica gel activation, 74
Silicon dioxide, porous, 28
Size exclusion, 2
Solid phase extraction, 2
Solubility of cyclodextrin, 86
  pH effect, 86
  urea addition, 86
Sorbic acid, 42
Spectrodensitometry, 105, 106, 107, 162, 165
Stainless steel wick, 159
Standard curve, ascorbic acid, 56, 117, 121
Substances, 18
Sulfated glucuronyl neolato-tetrosyl ceramide, 61
Surface areas, 16
Surface reactions, 15

Technicium, 148
Temperature effects, 30, 104
Temperature gradient, 81
Tetramethyl ammonium chloride, 42
Theobromine, 36
Theophylline, 36
Thermally labile compounds, 158
Thiobarbituric acid assay, 111
Thiogenal, 24
Triacylglycerol, 13
Two-dimensional electrophoresis, 187
Two-dimensional scanning for isotopes, 138
2-D process, 188
Tyrocidine, 36

Uridine, 33
Uridine-triphosphate, 33
Urine, 102

Validation, 69
Vanguard scanner, 139, 142, 147, 148
Video scanner, 185

Wettability, 20

Xanthine, 36

Zig-Zag, 186
Zip-Zag flying spot, 186

(*continued from front*)

Vol. 63. **Applied Electron Spectroscopy for Chemical Analysis.** Edited by Hassan Windawi and Floyd Ho

Vol. 64. **Analytical Aspects of Environmental Chemistry.** Edited by David F. S. Natusch and Philip K. Hopke

Vol. 65. **The Interpretation of Analytical Chemical Data by the Use of Cluster Analysis.** By D. Luc Massart and Leonard Kaufman

Vol. 66. **Solid Phase Biochemistry: Analytical and Synthetic Aspects.** Edited by William H. Scouten

Vol. 67. **An Introduction to Photoelectron Spectroscopy.** By Pradip K. Ghosh

Vol. 68. **Room Temperature Phosphorimetry for Chemical Analysis.** By Tuan Vo-Dinh

Vol. 69. **Potentiometry and Potentiometric Titrations.** By E. P. Serjeant

Vol. 70. **Design and Application of Process Analyzer Systems.** By Paul E. Mix

Vol. 71. **Analysis of Organic and Biological Surfaces.** Edited by Patrick Echlin

Vol. 72. **Small Bore Liquid Chromatography Columns: Their Properties and Uses.** Edited by Raymond P. W. Scott

Vol. 73. **Modern Methods of Particle Size Analysis.** Edited by Howard G. Barth

Vol. 74. **Auger Electron Spectroscopy.** By Michael Thompson, M. D. Baker, Alec Christie, and J. F. Tyson

Vol. 75. **Spot Test Analysis: Clinical, Environmental, Forensic and Geochemical Applications.** By Ervin Jungreis

Vol. 76. **Receptor Modeling in Environmental Chemistry.** By Philip K. Hopke

Vol. 77. **Molecular Luminescence Spectroscopy: Methods and Applications** (*in two parts*). Edited by Stephen G. Schulman

Vol. 78. **Inorganic Chromatographic Analysis.** Edited by John C. MacDonald

Vol. 79. **Analytical Solution Calorimetry.** Edited by J. K. Grime

Vol. 80. **Selected Methods of Trace Metal Analysis: Biological and Environmental Samples.** By Jon C. VanLoon

Vol. 81. **The Analysis of Extraterrestrial Materials.** By Isidore Adler

Vol. 82. **Chemometrics.** By Muhammad A. Sharaf, Deborah L. Illman, and Bruce R. Kowalski

Vol. 83. **Fourier Transform Infrared Spectrometry.** By Peter R. Griffiths and James A. de Haseth

Vol. 84. **Trace Analysis: Spectroscopic Methods for Molecules.** Edited by Gary Christian and James B. Callis

Vol. 85. **Ultratrace Analysis of Pharmaceuticals and Other Compounds of Interest.** Edited by S. Ahuja

Vol. 86. **Secondary Ion Mass Spectrometry: Basic Concepts, Instrumental Aspects, Applications and Trends.** By A. Benninghoven, F. G. Rüdenauer, and H. W. Werner

Vol. 87. **Analytical Applications of Lasers.** Edited by Edward H. Piepmeier

Vol. 88. **Applied Geochemical Analysis.** by C. O. Ingamells and F. F. Pitard

Vol. 89. **Detectors for Liquid Chromatography.** Edited by Edward S. Yeung

Vol. 90. **Inductively Coupled Plasma Emission Spectroscopy: Part I: Methodology, Instrumentation, and Performance; Part II: Applications and Fundamentals.** Edited by J. M. Boumans

Vol. 91. **Applications of New Mass Spectrometry Techniques in Pesticide Chemistry.** Edited by Joseph Rosen